AF768754

Forever Green

Plants in the dream time

Forever Green

Plants in the dream time

DAVE PEPLER
LIEN BOTHA & LAURIAN BROWN

H&R VISI

This book was published in
Human & Rousseau's
fiftieth anniversary year.

1959 - 2009

First edition, first impression 2009
by Human & Rousseau
an imprint of NB Publishers, 40 Heerengracht, Cape Town 8001

Copyright published edition Human & Rousseau (2009)
Copyright text and photograpghy VISI (2009)

All rights reserved
No part of this book may be reproduced or transmitted in any form or by any electronic or
mechanical means, including photocopying and recording, or by any other information storage or retrieval system,
without written permission in writing from the publisher.

Publisher: Ansie Kamffer
Editor VISI: Johan van Zyl
Proofreaders: Nelia Richter and Faan Herholdt
Translator: Peter van Noord
Designers: Anton Sassenberg and Gerhardt van Rooyen
Index: Anna Tanneberger

Reproduction by Virtual Colour, Cape Town
Printed and bound in China by WKT Company Limited

ISBN 978-0-7981-5094-1

Contents

Pharmacy in the Veld

I grew up in Robertson in the Little Karoo. My mother's family lived in the bodorp – Ouma Lilly, Oupa Boy, brothers and sisters – all committed gardeners and lovers of the veld. From them I learnt the ritual of "going into the veld" – filling a wine bottle with sweet coffee, wrapping it in brown paper and setting out at first light – always with a purpose: the first naeltjies of spring for the sitting room, kamkoo and vinkels for preserving, and num-num to eat on the spot. For me, though, the real treasure lay in the discovery of names like kukumakranka, renosterbos, geelpleisterbos, wild rosemary and mistletoe. I didn't know then that nearly all these plants contain hidden medicinal properties, but I was to learn that in South Africa, plants used as medicines number more than 4 000.

The use of wild plants as medication, elixirs, love potions and remedies is as old as humankind itself. One day, long before the written word, someone somewhere – out of need or plain curiosity – chewed a leaf, sniffed some scrapings of bark or nibbled a bulb and: Eureka! Relief, enchantment, euphoria – and they lived to tell the tale. Even in the very first inscriptions, such as the *Epic of Gilgamesh*, there are references to ceremonial and cleansing herbs:

Ninsun went into her living quarters.
She washed herself with the purity plant,
She donned a robe worthy of her body,
She donned jewels worthy of her chest,
She donned her sash, and put on her crown.
She sprinkled water from a bowl onto the ground.
– Tablet 3

Better known by its Hottentot name baroe, Cyphia volubilis *is a twining herb with small but showy flowers and a moisture-laden underground tuber that is traditionally used as a thirst-quencher and also as an emetic.*

The Bible speaks of balsam, bitter herbs, manna and rue. In the medieval manuscripts of the alchemists there are numerous references to powerful potions made from deadly plants. And the works of Shakespeare are steeped in references to elixirs and plant poisons, such as these lines from *Pericles*, Act III:

I have,
Together with my practice, made familiar
To me and to my aid the blest infusions
That dwell in vegetives, in metals, stones ...

Here in South Africa, Anders Sparrman, William Paterson and François Le Vaillant described medicinal plants long before Carl Peter Thunberg completed his great field study, packed with descriptions of herbs, for *Flora Capensis* in 1813.

Today, a healthy kinship exists between traditional herbalists, western homeopathy and oriental influences from Ayurvedic treatments. Almost every day we hear of new preparations making news worldwide, such as the cancer bush, the African potato and devil's claw. You have only to look at the shelves laden with herbal preparations to see that there are remedies for everything from ingrown toenails to sties.

Not all plant-derived medicines are beneficial, nor are they necessarily dangerous. We tend to underestimate the power of the placebo; blind faith in a preparation can have remarkable results. Rue, for instance.

A year or two ago I asked my neighbour for a slip. "With pleasure," she said, "but remember, rue works only when it is given, not when it is asked for." Last year I heard the same story from a Mapuche Indian in Patagonia. We don't realise how deeply our different worlds are intertwined by this ancient, universal knowledge of plants. The moral of the story: don't wait to be asked – give!

THE HEALING POWER OF NATURE

Only in a laboratory can the medicinal powers of plants be separated from their perceived magic. In the ancient world, prayer and prescription went hand in hand. Though not all sacred herbs may have had pharmacological value, over millennia a vast bank of plant knowledge accumulated, interwoven with faith and superstition.

Scientists have drawn on this knowledge to produce important medicines such as digitalis, morphine, quinine, curare and reserpine. Other plant contributions include aspirin from meadowsweet, a breakthrough cancer treatment from the Madagascan periwinkle, and, from the Mexican yam, the raw material for oral contraceptives.

In South Africa, it is estimated that 70% of the population relies on plants for primary health care. There are also an increasing number who depend on these plants for their livelihood. In the past, harvesting of medicinal plant material was governed by taboos and undertaken only by

---

Clockwise from top left:
Dodonea angustifolia: *A decoction of the leaves is used for fever, colds and arthritis.*
Cyclopia genistoides: *Used to make an antioxidant, caffeine-free herbal tea.*
Melianthus major: *Leaves are used in poultices and decoctions.*
Bulbine frutescens: *The sap is used for burns and all kinds of skin problems.*

skilled traditional healers. Demographic and economic pressures have brought in other gatherers; tons of bulbs, tree bark and plants arrive at muti markets every week, causing numerous plant species to suffer serious decline.

The management of this phenomenon has presented a huge challenge to conservationists and health regulators. Collaboration with communities to teach correct harvesting practice and the cultivation of muti plants is aimed at addressing the situation, as is the registration of traditional healers. The Institute for African Traditional Medicines has been set up to evaluate the effectiveness and safety of traditional medicines and to provide consumer protection.

Key medicinal substances are usually the plant's defence against predation and are often pretty potent. "Only the right dose differentiates a poison and remedy," declared Paracelsus. A number of kill-or-cure plants feature in the traditional African pharmacopoeia, and probably also in your garden. Consider your clivias: looking at them is all you really need do to feel better, but those pretty faces are also awesome reminders of the power of nature.

BENEFICIAL PLANT BASICS

- Classic herbs often contain beneficial compounds. Grow and use them lavishly.
- Salad leaves: Grow a mix of bitter (chicory, endive); peppery (rocket, mustard and cress); sour (sorrel); salty (our own delicious brakslaai, *Mesembryanthemum crystallinum*); and simply succulent (butter lettuce).
- Fragrant aromatic plants such as scented-leaf pelargoniums, to crush and inhale after a carbon-overloaded day.
- Tea: Ceylon tea (*Camellia sinensis*) is one of the greatest medicinal herbs of all time but, like rooibos, not for home cultivation. Try lemon verbena (*Aloysius triphylla*) or indigenous *Lippia javanica* for a delicious lemon-flavoured tea good for treating colds.
- Plants to steep in brandy: Buchu (*Agathosma betulina*) is a great South African cure-all and a lovely fragrant plant for the patio if your garden doesn't have the acidic soil it prefers.
- Soothing gels: The sap of *Bulbine frutescens* is a healing gel, good for burns and all kinds of skin problems. Grow it on a sunny windowsill in the kitchen. It also makes a wonderful ground cover in hot, dry conditions.
- Sovereign remedy: Shimmering blue rue (*Ruta graveolens*) is an ancient, sacred and protective herb and also a great traditional panacea for fever, rheumatic pain, hypertension and venous disease. It is also the herb of grace – and so a must for every garden.

The renowned cancer bush Lessertia frutescens *is an old Cape cure-all for stomach problems, internal cancers, fever and diabetes. It is rich in amino acids which may boost the immune system and act as a general tonic.*

Cathedrals of the Veld

The crowns of trees are the domain of birds, children, the mad and the holy.

Like my great-aunt Ella, spinster of Robertson. She and her sister, Susie, lived in a dark house next door to my grandmother Lilly. Ella played the organ for the Salvation Army and drove Ouma to distraction by coming over daily to peer into the pots on her stove. One Sunday, Ouma sent me over to borrow an onion from Aunt Ella. I walked into the gloomy hall and through the kitchen, but Ella was nowhere to be seen. From the back stoep I shouted, "Coo-eee", and from the depths of the garden there came an answering "Coo-eee". There I found her – in the topmost branches of a naartjie tree, baby sparrow in hand. She was about to restore it to its nest. Tree-climbing was Ella's forte. The sight of her bloomers under a knitted brown dress haunts me still.

I suspect that humankind was able to cleave stone before it learnt to cleave wood. That moment when we first exposed the hidden grain was surely one of our greatest intellectual leaps, opening a path to discovery. Go and reread Jacob Bronowski's lyrical essays on the subject in *The Ascent of Man*.

Trees take us back to the real beginning of our humanity. Because the thinking being seeks the universal and finds something of it in geometry, trees are something of an enigma. Their design and architecture appear totally chaotic. Lie on your back against a tree trunk and look up. The only recognisable form is the shadow line, which is usually round, but the growth of the trunk and branches is wild. And yet, if you let your eye wander out of focus, patterns and designs gradually begin to emerge from the ostensible tangle. South Africa's indigenous trees are among the most beautiful in the world.

The wild laburnum, Calpurnia aurea*, is a small, fast-growing tree, making it a superb choice for small city gardens. Native to our evergreen and riverine forests, it blooms in the heat of summer, producing cascades of yellow pea flowers.*

The baobab, the flat-crowned acacias of the Bushveld, the quiver trees of the Richtersveld: all are a stylist's dream. With or without leaves, they remain heavenly designs, small cathedrals of the veld. These are a few of my favourite trees: *Acacia karroo*, or the sweet thorn, is Africa's great all-rounder. The leaves are a favourite nibble of many kinds of game, the scent of those butter-yellow or cream flowers is heady, hordes of butterflies depend on the tree, the sweet gum …

Then there is *Boscia albitrunca var. albitrunca*, or the shepherd's tree (also in the basic starting list below). Kudus keep pruning this tree, which happens to be protected, throughout its long life, leaving us with a sculpturesque tree which defines a landscape. The grey-white trunk with its rounded canopy forms the perfect resting spot. When I go hunting in Namibia, I always eat an orange in the shade of one of these trees, to affirm that I am there.

The Namaqua fig or rock breaker, *Ficus cordata*, must surely be one of our most determined trees. Drive north of Garies and there, against the bare cheek of a rock slab, you'll find it growing obstinately. The roots coil like snakes over and under the stone and the only redwing starlings for miles around call from among the leaves.

Last but not least is the wild olive, *Olea europaea subsp. Africana*. Plant this icon in your garden and wait for autumn, when the little purple fruits begin to fall. Out of nowhere, bulbuls, white-eyes and the yellow-billed bush dove, that phantom of the undergrowth, will suddenly appear. You'll have birdsong at your window, shade against the late summer sun and a tree that stands forever.

Could there be a better way to keep a child quiet than to let him climb a tree or hide away among its branches and leaves? Where might you find a sounder sleep than under the shadow of a willow next to a lazy stream? Could you feel more liberated than when you walk in the green depths of a damp forest? Plant an indigenous tree and climb into it when it's grown. Only then will you be truly grounded on this earth. And able to talk, like cultural ecologist David Abram, of "the spell of the sensuous".

ROLL OF HONOUR

South Africa's list of national trees is a long one. The first comprehensive survey, undertaken in 1965, consisted of 1 009 species, subspecies and varieties. In 1986 the Dendrological Foundation published a booklet listing 1 073 numbered species (classified under 370 genera and 97 families). The basic selection of five of our numbered beauties below should inspire you to start planting indigenous.

- Shepherd's tree (*Boscia albitrunca var. albitrunca*) – Found in the dry regions of the country, where its green umbrella provides shelter for man and beast. This small, dense, evergreen tree bears starry, fragrant flowers in spring, followed by yellow fruits that provide a feast for veld dwellers. In the old days of trekking, the roots served as a substitute for coffee. It can handle brack soil, drought and mild frost.
- White stinkwood (*Celtis africana*) – Thrives in a variety of conditions. It is large with a spreading crown. Plant it in good soil, give it enough food and water and it'll grow up

---

Clockwise from top left:
Crimson joy: the weeping boer-bean, Schotia brachypetala.
African marble: the lichened bark of the sweet thorn, Acacia karroo *(also bottom left).*
Green mansion: the wild plum, Harpephyllum caffrum.

to two metres a year. In winter the leaves fall to show off its silvery white trunk. It is reasonably resistant to drought and frost.

- Wild olive (*Olea europaea subsp. africana*) – A small to medium-sized tree that does well in a variety of regions. Resistant to frost and drought.
- Henkel's yellowwood (*Podocarpus henkelii*) – One of the most beautiful and popular of the yellowwood species. It is an evergreen conifer that can grow to a noble height of 30 metres. If it receives enough light from all sides, the rippling foliage can hang all the way down to the ground. This makes it a wonderful windbreak. The golden-brown bark is also attractive, peeling back in large strips to reveal satin-soft russet skin beneath. It occurs naturally only in the damp forests of the Eastern Cape, Transkei and KwaZulu-Natal, but will grow in both sandy and loamy soil if there is enough moisture. Reasonably resistant to drought and frost but is a slow grower.
- The weeping boer-bean (*Schotia brachypetala*) – Bears trusses of striking carmine flowers, which appear on the old wood in spring or early summer and attract masses of birds and insects. Semi-deciduous, it's a good choice for the smaller garden because it grows no more than about ten metres high. The young leaves are bronze-tinted, changing to pale and then leaf green. As winter approaches, large, flat, brown pods appear. The boer-bean occurs countrywide, especially in warm and dry habitats. A slow grower; drought-resistant but frost-tender while it is young.

TREE PLANTING BASICS

- Choose the site and the kind of tree with extreme care. Get expert advice and find out what its mature size will be. Plant a tree that will ultimately outgrow its site and you condemn it to death at the outset.
- Work out the position of the sun through the seasons and where the shade will fall in summer and winter.
- Make sure that the tree will not eventually grow too close to walls or buildings and that its canopy will not extend over into the next-door garden. Neighbours are notoriously tricky when it comes to trees.
- Remember that the height and proximity of surrounding buildings will affect the growth habit of the tree, drawing it upward.
- Large trees are available from specialist growers, but small saplings, well planted and watered, settle in and grow more quickly.
- Autumn and winter are good planting times because the roots have time to settle in during the dormant season when the soil is cool.
- Shape your tree from its earliest stages. Superfluous branches are best removed when no more than 1 cm thick.

Clockwise from top left:
Pure gold: the wild laburnum, Calpurnia aurea.
Handsomest evergreen: Henkel's yellowwood, Podocarpus henkelii.
Defining scarlet: the coastal coral tree, Erythrina caffra.
A picture in autumn: the Transvaal privet, Galpinia transvaalica.

Gifts from the Earth

I can still hear the north wind blowing against the kitchen window. Then Oupa would say: "There goes the steam train whistling up Goreeshoogte; it will snow tonight." The ice-cold house was dark, except for the kitchen, where we sat around the Husqvarna stove. My father, with a newspaper on his lap, my mother with her eternal piece of crochet work, and us children dead quiet, listening to the wireless on the grocery cupboard.

Just before the nine o'clock news my mother would reach for her oven mitts, open the oven door, draw out a steel pan and place it on the table. We would each get a sweet potato on a little wooden board. Beneath the broken skin, tough as toffee, under a layer of dark caramel, was the pale, soft yellow of the potato, sweet and smoky, with a faint smell of earth. Wintersweet potato, from our own garden.

Our town had large plots, big enough for a formal flower garden bordering the street, a vegetable garden, an orchard and sometimes even a fowl run. Gardening was part of the local culture, and cause for pride and envy. The town still received irrigation water via furrows in the street, and nothing on earth can set gardeners at one another's throats more surely than the scrabble for water. In high summer, at midday, the only people on the street were those leading water, barefoot, trousers rolled up, spade over the shoulder, and a sack or folded newspaper to seal the sluices.

When your turn came, everyone stood at the ready, because you had to stop the furrows breaching or your water from disappearing suddenly down a mole hole – the water had to be led. The hot soil bubbled and sizzled when the water flowed over it, and you walked along with the wet flow, the mud between your toes.

Gardeners are truly earthed.

In a vegetable garden each season has its echo. Winter is my favourite time, because the cold earth rests and waits; deep in the storeroom a basket with slumbering seed potatoes lies in wait,

Exotic, symbolic flowers, simply delicious fruit; the passion vine, Passiflora caerulea *feeds both body and soul. The most commonly grown species is the granadilla,* P. edulis, *with its wrinkled, dusky purple fruits.*

and the pruned trees stand black and stark against a weak sun. In the early morning you see the compost heap smoking like a small volcano and the only movement is an olive thrush searching for earthworms among the wet leaves.

And how do you know when it's spring? One Saturday morning your mother begins to pack out jars of seed: lazy housewife beans, radish, beetroot, boer beans. The seed beds have their little flags at the head of neat rows and the large beds get compost and manure. Mornings, coffee mug in hand, you wait for the first seed to show. Then it all gets very busy, with weeding, thinning of radishes, turnips and carrots, the pulling up of potatoes – and suddenly there are honey-sweet baby peas to steal while you work.

Then everything ripens together. Your mother sends you to Tant Nellie with a picking of beans. Tant Nellie gives you four caramel-brown hen's eggs in return, and asks you to drop off a few squash for Tant Martjie on the way home. Tant Martjie puts a jar of curried beans in the basket and sends you next door with a giant beefsteak tomato for Tant Fifi. How rich we were, because these baskets with gifts were our common bond with the soil.

An authentic food culture is established through simplicity. Is there anything more elegant than a meal that consists only of an earthenware dish full of freshly dug new potatoes, cooked in their skins with a sprig of rosemary, with nothing more than farm butter, coarse salt and ground pepper?

Every day I hanker after a vegetable garden, because I want to learn to wait again, to watch how seeds become plants, and then, with reverence, to receive the gifts of the earth in a communion with the soil.

GARDEN TO KITCHEN

Growing your own vegetables makes all the difference to the way you cook and eat. It also means that you know exactly what has gone into and onto what you're eating and how far it's travelled. Food miles aside, there's nothing to equal a freshly picked green salad, a tomato with the bloom still fresh on it, or baby green beans that have come straight from plot to pot.

Self-sufficiency is ideal, but demands space and dedication. For the average family to be fully self-sufficient year round, vegetable guru Joy Larkcom recommends a ten by ten metre plot. Most gardeners today, even if they had that space to set aside, couldn't spare the time, the labour or the water involved.

The great thing is to grow and harvest and enjoy, and to do that, all you need is a bed of about one metre by two metres – a patch for pure pleasure.

If all you have is a sunny balcony or a courtyard, containers can also provide that pleasure. They should be at least 45 cm deep, with a richly composted soil mix. Add a spadeful of manure to the bottom layer and feed with a slow-release granular organic fertiliser.

It's also not absolutely necessary to set aside a special patch for vegetables. They can look wonderful among your flowers, provided they have enough sun and breathing space.

Clockwise from top left:
Swiss chard "Bright Lights": colourful cut-and-come-again staple.
Aubergines: one of late summer's many mellow pleasures.
Zucchini flowers and fruit: a gardening gourmet's delight.
"Blue Peter" runner beans: as decorative as they are delicious.

WHAT TO GROW

Your first thought will be for your favourites, but any vegetable is worth trying, so have a look through the seed racks at the nursery and see what takes your fancy.

Check the days to harvest time, always stated on the back of the packet. Quickly maturing vegetables like leaf greens – Swiss chard and oak leaf lettuces – make the most of a confined space. You snip off leaves as you need them and the plants continue growing.

Salad greens that need to be picked as fresh as possible are another top priority.

It's important to plant the right vegetables and varieties at the right time. A number of vegetables, such as lettuces and cabbages, have both cool season and warmer season varieties.

Getting to know varieties is half the fun – which are the most flavoursome, handsome, compact or quick-maturing. If you're buying seedlings, make sure the cultivar is specified on the label.

Nursery seedlings are less time-consuming and a good way to start. But growing your own seed offers a much greater variety and the whole process is magical and much more satisfying. Obviously, with things that need to be sown directly – such as beans, radishes and sweetcorn, there is no other way.

There's no point in growing vegetables without herbs – annual and perennial – so don't forget to make space for those.

KITCHEN GARDEN BASICS

- Vegetables need good sunlight and an absolute minimum of six hours of sunlight a day.
- Fertile soil is essential. Double digging and the incorporation of manure plus loads of compost will repay you handsomely in the long term.
- All vegetables need regular watering for steady, even growth.
- A daily check for pests or signs of plant distress will help to nip problems in the bud.
- Seedlings in particular need protection from slugs, snails, birds, cats, dogs, and a too-hot sun. Suit the protection to the problem: snail traps, cages, netting, shade cloth or tunnels.
- There's no point in growing your own vegetables if you're going to spray poison all over them. Getting the soil right first makes for long-term gardening health. Chemical fertilisers promote lush, soft growth, which is a magnet for pests. A watchful eye is your first line of defence. Familiarise yourself with home remedies as well as new organic ranges of pest control measures.

Clockwise from top left:
Rue, great cure-all and the herb of grace, is a must for every garden.
Fennel is essential with fish – and a magnet for beneficial insects.
The flowers of the onion family are a decorative bonus.
Handsome rewards: Globe artichokes provide spring's finest delicacy.

Stoep Story

I drag a stool to the low street wall and climb onto it. Standing on tiptoe and craning my neck, I can just reach the lowest bunch of wax flowers, with my tongue. They opened this week. At the heart of each flower, a round drop of nectar gleams. I close my eyes and let the nectar trickle into my mouth. When I open them again, I see the street through a lens of leaves.

My Ouma Lilly's stoep in the bodorp was a place for people and plants, a street salon where you could converse, read, daydream or doze, but above all, observe. It was our camera obscura on the street.

However, the stoep as a place to meet, and live, has long been a thing of the past. With the death of radio and the birth of television, and the arrival of the affordable car, the stoep lost its social function.

Most of us have forgotten our former street culture: when families would go for an evening stroll, like the Spanish paseo; when we would take baskets of garden produce to our neighbours; when the front stoep and its plants were the pride and joy of the lady of the house; when you could sleep out on a stoep bed and wake in the early hours to the whistle of a steam train, the play of phantoms from the streetlamps on the wall.

I am convinced the plant-filled stoep evolved in hot climates. Think of a narrow street in Seville: through the bars of the gate you can see the portal where a widow in a jet-black dress sits watching the street, the shadowy placita beyond, the light-green leaves of pot plants against the patterned Moorish tiles.

Think of Istanbul in summer, where the narrow streets have driven the verandas upwards and turned them into balconies – you are constantly aware of eyes behind the plants, seraglio eyes heavily outlined in kohl. In Troisk, a little town in Kazakhstan on the old Silk Route, I remember how, with a fat forearm, an old babushka parted a screen of salmon-coloured hollyhocks to peer out into the dusty street.

Codiaeum variegatum, *from the jungles of Malaysia, has the most colourful of all leaves. A range of dazzling cultivars are available in stained-glass colours from bright red through orange and yellow to purple. Space and strong light, with some sun, show the colours at their brightest.*

How predictable stoep gardens are – philodendrons, cinerarias, crotons, a ficus, a maidenhair fern on a rickety wooden stand – and cyclamens, those plants with beautiful leaves and extraordinary flowers. Stoep plants must like coolness and shade, tannies who polish their leaves and smokers who stub out sneaked cigarettes in their pots.

Now we have replaced our stoeps with balconies and atriums; we no longer watch the street or chat with the neighbours. But we do still have the quiet presence of plants, which is better than nothing because, even in their pot prisons, plants connect us to the faraway world of the soil.

Happily, near where I live, the stoep is still alive and well. Where I ride my bicycle through Ida's Valley, every house still has one, complete with plants and a tannie. The floors are painted red or green and there are pots with plants in them placed brazenly all around, or festooning diamond trellises. A grandmother, cocooned in a blanket and woolly hat, peers at the street through a curtain of leaves. People still come over to greet you and when you stop, you get all the news of the street. Ask for a slip of a special plant and it will be given with pride. Only in this kind of community are there still stoeps – stoeps with people and stories and plants. This is the kind of street where I will always want to live.

THE RETURN OF THE EXOTIC

The majority of our classic stoep and indoor plants are natives of tropical forests. They thrive in warmth, humidity and low light, which is what makes them able to survive under cover as well as indoors. After years of scorn and neglect, these lush creatures are back in fashion and with good reason.

Firstly, as cities burgeon, we long for the wild, and tropical plants are as wild as you can get. Secondly, as waterwise leaves spread their grey dapple over gardens (beautiful and sensible as that may be), we feel another longing, this time for the vast umbrellas and deep greens of the jungle. Suddenly, we see these familiar yet forgotten exotics with a fresh eye and new pleasure. Lastly, they serve to remind us not only of the great forests under threat worldwide, but also of a gentler habitat already lost – the stoep itself.

So, it is time to give these exotic creatures the limelight they deserve in your decor. On stoeps and in indoor garden arrangements, they have traditionally been placed together in tight groupings. The trend now is to feature a single plant or geometrically spaced groupings in a clear space or minimal setting. Placing them close together creates more humidity and more favourable growing conditions, so it is important to make sure that you provide adequate humidity for solo plants.

Treat each plant as a sculpture. Invest in a beautiful container that suits its shape and colour. Look for – or create – elegant frames for plants such as philodendrons, that require climbing support.

All plants need more than bare soil around them to look their best. Larger pots and plants can be underplanted with ground covers, or even shade grass. Moss, bark or pebbles make good finishes for jungle plants.

Clockwise from top left:
"New Guinea" hybrid Impatiens *are large, lush cousins of the beloved Busy Lizzie.*
Graceful and strikingly coloured Dracaena marginata tricolor *is a native of Madagascar.*
Fittonias *are mosaic-leaved little creepers from the rain forests of Peru.*
Scindapsus *"Marble Queen" is an easy-to-grow climber from the Solomon Islands.*

CLASSIC HOUSE PLANTS

- *Adiantum capillus-veneris* (black maidenhair fern)
- *Aglaonema crispum* "Silver King" (painted drop tongue)
- *Alocasia x amazonica* (Amazon lily)
- *Codiaeum variegatum* (croton)
- *Dieffenbachia seguine* (dumb cane)
- *Dracaena marginata* (Madagascar dragon tree)
- *Hoya lanceolata* (wax plant)
- *Maranta leuconeura* (prayer plant)
- *Monstera deliciosa* (delicious monster)
- *Philodendron*

HOUSE PLANT BASICS

- A draught-free spot with bright light but no direct sunlight, which may scorch the leaves.
- Water carefully: keep them moist but don't drown them.
- Stand the pot in a bed of pebbles in a tray of water. Also give the leaves a daily misting with tepid water.
- Remove dead leaves and flowers and clean the leaves regularly.
- Repot plants annually.
- Nip pests in the bud. Keep a lookout for scale and mealie bug, the most common pests.
- Feed plants regularly, or use a slow-release granular fertiliser.
- Give plants a rest period, usually in winter when feeding should be stopped and watering reduced.
- Invest in a good book in which you can check the specific needs of your plants: *House Plants* by Paul Williams (Dorling Kindersley, 2006) contains excellent photographs that show the shape and detail of plants, as well as all you need to know about growing and positioning them.
- A number of house plants function as air fresheners. These include *Ficus benjamina* (weeping fig), *Spathiphyllum wallisii* (white sails), *Calathea makoyana* (peacock plant) and *Chlorophytum comosum* (hen-and-chickens).

Philodendrons, kings of the stoep jungle, are native to the tropical forests of America. There are both climbing and non-climbing species, with leaves that range from heart-shaped to palmate. Philodendron *"Xanadu" is one of the very best non-climbers.*

Fynbos Fantasy

The Cape of Good Hope is much more deserving of the name Tierra del Fuego than that group of sodden islands half a world away. In this astonishing cape and its interior is one of the earth's richest flower gardens, a garden formed by fire.

Fynbos has no trees and is full of tough, thorny plants; from a distance it looks drab and monotonous. But walk into the veld in spring and this drabness is revealed as a world of wonders – as you walk, stipples of colour and texture catch your eye, like the heart of an opal.

Most of the flowers are small and delicate, like the orchids and the gladioli, but here and there big, bizarre inflorescences lift their gleaming heads. Proteas and pincushions are so large and colourful, so abstract, that they appear unreal.

Take a fresh look at the king protea (*Protea cynaroides*), however. All my life this flower has irritated me with its pink pink, its perfect symmetry – not a hair, not a petal out of place. No wonder it's the flower of church decorations and desperate tourists wandering through the late-night lukewarmth of departure lounges. But one must also remember that familiarity is one of the greatest obstacles to accurate observation. Rather look again and again, because only then can the protea's immediate vulgarity be transcended and its real fascination revealed.

Explore the group further and look at the more modest species like *P. nana* and the low-growing ground proteas – species such as *P. acaulos*. After a lifetime of climbing, I know that among the greatest rewards that a Cape mountain can give you is an unexpected rainbow above a waterfall, the chemical shock of crushed buchu, and the wine-red of a mountain rose with the light behind it.

Pincushions and leucadendrons are the more modest sisters of the Protea family. It's no wonder that the flower-pickers describe them as "tints". They are not nearly as flashy as the proteas;

Ambassador extraordinary: A unique and startling combination of size, shape and symmetry make Protea cynaroides *a worthy national flower for South Africa. The spectacular flowers may be as much as 30 cm across.*

they are smaller, more delicate and with colours which shade subtly into one another. Of these two groups the silver tree is the most spectacular, with its silky leaf covering that shines like fur on an animal. And what about *Leucadendron salignum* with its petals that darken from farm-butter yellow to henna? Even in death, after a devastating veld fire, they remain lovely, with the calices as delicate as puff pastry still on each blackened tip.

The protea and its extended family share an exquisite visitor, a living ornament. Go walking in the Kogelberg and seek out a patch of sugar bush in full bloom. Find yourself a spot of shade and wait, and soon you'll hear the first duets, loud, shrill and unmelodious. A pair of Cape Sugarbirds feeding in their territory, the males perched cheekily in the tallest flowers, dashing tails bent in the southeaster. Then, a flash of metallic green – the Malachite Sunbird. Then gold and orange – the Orange Breasted Sunbird. Then green and red, braided with blue – the Southern (Lesser) Double-collared Sunbird. All of the crown jewels of the fynbos.

So my irritation with proteas, pincushions and leucadendrons has now changed to admiration. After years of looking at them, their colours and shapes now make sense. With their satin pink, silver, chrome-yellow and terracotta, the meticulous overlapping of the petals, these indigenous designs are worthy of adorning the Art Deco icons of Madison Square Garden. Proteas in the foyer of Radio City Music Hall and on the grillework of the New Yorker Hotel? Definitely.

FAUNA-LIKE FLORA

The roots of the Protea family lie in ancient Gondwanaland; continental drift has taken them as far as Australia, Asia and South America. The South African proteas embrace nine genera and over 400 species. While the vast majority grow in the fynbos of the winter rainfall regions, there are proteas all through the Southern and Eastern Cape and even as far north as the Limpopo River. They range from huge billowing shrubs and smallish trees to dainty little bushes and ground-huggers, all with unflowerlike flowers characterised by a mesmerising radial symmetry. Fur and feathers, antenna-like stamens and pangolin plating make them seem more like fauna than flora. And they are not easily tamed.

PROTEAS IN THE GARDEN

A great deal of research has been done, particularly at Elsenberg Agricultural College, into cultivating and developing more adaptable and resilient – as well as even showier – cultivars for both professional growers and the home gardener, with great success. But the crucial factor still lies in selecting the right species or cultivar for your climate and especially your soil, rather than trying to change your soil to suit the proteas you want to grow.

There's another wild aspect to our furred and feathered friends and that is in their uncompromising style. Proteas are social, informal plants, best suited to gardens of artful wildness. They look and grow best in groups and drifts of different types – as they grow in the wild, combined with other fynbos plants, such as buchus and restios. Study your local species in the wild to see

---

Clockwise from top left:
The common and beloved sugar bush, Protea repens, *blooms nearly all year round.*
Protea *"Sylvia" flowers from November to February.*
Protea cynaroides *usually begins flowering in April.*
Leucadendron *"Safari Sunset" may be harvested from autumn to spring.*

how and where they grow and with what other plants, before you plan your planting. Plantings at Kirstenbosch will provide superb inspiration for Western Cape gardeners.

Fynbos plants are on the whole relatively short-lived. Careful selection and breeding have succeeded in extending their life spans, but after five or ten years they may still become gaunt and leggy. In the wild, fire sorts this out, combining clearing with regeneration, but this is not an option in the suburbs. Fynbos fundis advise clearing the lot and starting all over again, because it's best for all the plants to grow up together, just as they would after a fire.

Seed should be sown in autumn, in a sunny situation, in a very free-draining medium. Kirstenbosch recommends two parts coarse sand, one part leaf mould or milled bark, and one part loam. Most plants flower only after five years.

Tip cuttings should be taken from a strong stem in late winter or spring. They will take about three months to root.

PROTEA BASICS

- A sunny aspect, good drainage, plenty of sun and good air circulation.
- A combination of regular watering and perfect drainage. Proteas are often found growing in areas of seepage in the wild.
- Most species prefer acid soil, but there are coastal species which will grow in alkaline soils with a pH of up to 8.
- No manure or phosphates, but you can feed with nitrogen and potassium and ammonium sulphate.
- Roots must not be disturbed and should be kept cool and protected with a thick layer of compost or pine bark mulch.
- Start with small plants and pinch out the central growing tip on planting. Continue to pinch back to encourage bushiness.
- Plants with a mature height of less than two metres should be planted 0,65 metres apart. Species over two metres should be planted one metre apart.
- Regular deadheading and pruning will keep the plant bushy and more resistant to wind – and also prolong the life of the plant.

In nature Protea cynaroides *grows in sandy acid soil, usually in damp areas, but in cultivation it has proved surprisingly adaptable and is one of the most rewarding species to grow.*

Staying Power

They are die-hards that seem to have been created with church arrangements and funeral wreaths in mind. Week after week, pincushions retain their freshness and colour. Eventually, out of sheer exasperation at their continued perfection, you throw them out and replace them with a new bunch. (Those perennial arrangements on the reception counters of estate agents and private hospitals also fall into this category.)

There is something awesome about the genus *Leucospermum*, or pincushion. Like all plants, they can't flee from danger but must stand their ground. Fynbos burns regularly, so to survive fire is an issue – unless you establish special defences. This is how it works: to remain vital, fynbos must burn every 12 to 15 years. Fire opens the canopy, allowing light to hit the ground and giving bulbs the chance to flower again – until the canopy closes in once more. Most of the larger fynbos plants retain their seeds in the seed bud; only after exposure to fire will it open – almost instantaneously – to release thousands of seeds.

For an organism that is trying to disperse its genetic material effectively, the next issue is that the seed might fall mostly around the mother plant. Two possible solutions present themselves: One, make your seed as airworthy, as light and feathery as possible, so that the wind will do the job. Two, pay someone else to carry it away for you. This makes *Anaplolepis steingroeveri* and *Pheidole capensis* the most important ants in the fynbos. In all similar regions of the world – California, Australia, Chile and the Mediterranean – notable partnerships have developed between plants and ants. Same here.

The seed of *Leucospermum*, like that of proteas, stays in the seed capsule until released by fire, when it falls plumb to the ground. What makes it different is that each seed has a small gland, the elaiosome, the scent of which is irresistible to indigenous ants – and mice and rats.

The flowers of Leucospermum cordifolium *vary in colour from yellow to salmon, pink, vivid orange and crimson. They are borne on a handsome spreading shrub that grows to 1,5 metres tall and two metres across.*

So begins an armaments race between ant and rodent. The ants collect the seed, as ants will. They carry it off and down into their nests, where only the elaiosome is eaten. There could hardly be a better scenario for germination: It's summer (when most veld fires occur), the ants are at their most active, the temperature is perfect for germination and the seed is already underground, where moisture levels are higher and the soil has been enriched with minerals leached from the fresh ash after the fire. Definitely a head start, compared with other seeds that must lie on the surface and wait. So when next you walk past an arrangement of *Leucospermum*, pause in admiration. Then give a moment's thought to their little collaborator, the humble ant.

GRANNY'S SEWING BASKET

It's not difficult to identify a *Leucospermum*. They resemble nothing so much as the colourful pincushion in your grandmother's sewing basket, only much neater. The long stamens have knob-like stigmas on the tips and stick out like pins from the flower head. These striking flowers are borne on handsome evergreen shrubs that are probably the most attractive in the Protea family. Form and pattern and flower make them winners in any fynbos or Mediterranean-style garden. They are excellent for picking and their sweet nectar is also a magnet for sunbirds.

PINCUSHIONS IN THE GARDEN

There are about 50 species of *Leucospermum*, most of which flower between June and November and range in height from 0,6 metres to three metres. In addition to the species, there are numerous excellent and versatile cultivars, at least one or more of which will be right for most gardens. Some have been bred with long stems for the cut flower industry. There are even one or two, such as *L.* "Highgold", that will tolerate alkaline soil.

- *Leucospermum cordifolium*, a rounded, spreading shrub that grows to two metres, is one of the most vigorous and attractive species, with orange and yellow flowers in summer. There are also a number of lovely *L. cordifolium* hybrids and cultivars, including "Yellow Bird", which has pink-tinged flowers.
- *Leucospermum* "Caroline", one of several beautiful crosses between *L. cordifolium* and *Leucospermum tottum*, grows to 1,5 metres and blooms in early summer.
- *Leucospermum formosum* has grey leaves, grows to three metres and bears yellow catherine-wheel flowers in spring.
- *Leucospermum mundi* has yellow and orange flowers and is suitable for small gardens.
- *Leucospermum reflexum* is a tall, very effective species with grey leaves and rocket-like flowers, and is a superb backdrop or screening plant.

Professor Kristo Pienaar regarded *L. cordifolium* and *L. reflexum* as the best garden species, but also recommended *L. patersonii*, *L. tottum* and tree-like *L. conocarpodendron,* at five metres the tallest of the species.

The tufted pincushion, Leucospermum oleifolium, *has flowers which turn from yellow to orange and crimson as they mature, making this compact shrub a dazzling sight when in bloom.*

PINCUSHION BASICS

- Pincushions occur naturally in the winter rainfall region of the Southwestern Cape, so the plants must receive sufficient water in winter.
- In the wild they usually grow in full sun on mountain slopes in well drained sandy or stony ground. Favourable growing conditions can therefore be created by planting them in an open, airy position in fairly poor, slightly acid soil to which no fertiliser has been added.
- Plant pincushions in March and September in the winter rainfall regions. In the interior, September is the best time.
- Make the hole slightly bigger than the bag in which you bought the plant. Remove the bag very carefully and try not to damage the surface roots. Press the plant in lightly but firmly and cover the surrounding soil with a thick layer of pine bark, peat or compost.
- Water regularly and thoroughly. Do not disturb the ground around the shrub, because the surface roots are very sensitive. Any weeds should be removed by hand.

For more information on growing pincushions, turn to the chapter on proteas on page 30.

Clockwise from top left:
Leucospermum reflexum *has styles which reflex outwards as the flower matures, borne on a handsome shrub with blue grey foliage.*
The styles of Leucospermum cordifolium *flowers are curved inwards, gradually unfurling as the flower matures.*
A fading flower of Leucospermum reflexum, *showing the styles completely reflexed.*
Leucospermum tottum *has large pale flowers with darker tips and is deservedly known as the elegant pincushion.*

Honey Bunch

There is a stone house at Suurvlakte, in the hidden valley beyond Wolseley. To get there, you must drive over the pass known to locals as the Dom-en-Long (years later I found out the story behind the name: the engineering firm that had erected the power lines along the old forestry road was Dorman, Long, Swann and Hunter – a twist too far for the Cape tongue).

It was here, under an ancient rockwood, that I would come to take a break in the heat of midday. One hellishly hot February day, I fell asleep under my tree. I dreamed that I was struggling, an insect trapped in a pool of honey, and woke up with a jolt. The explanation was simple: the patch of pink ericas around me had dispersed their rich honey scent across the veld.

Fourteen species of *Erica* occur in Western Europe, two in Ethiopia, nine in East Africa and 688 in South Africa. Of those 688 species, 666 are endemic (that is, they are found only here), 136 are on the Red Data list and 31 are endangered species. In the Caledon district alone more than 220 species can be found.

You may well wonder why so many in a region that must have just about the poorest soil on earth: pure, deep-leached sand, high acidity and virtually no nutrients or trace elements. Summers are boiling hot and winters wet and cold – yet here the ericas are at their most spectacular. This might even be the locus where the astonishing variety of species originated and spread.

But the secret of the erica's survival lies in the alliances it has established with other forms of life. This plant is a skilled courtesan, a seductress from tip to toe (or rather root). A partnership with a benign root fungus ensures that nitrogen in the air is made accessible to the plant, but it is the symbiosis with birds and insects, especially hawk moths, that is particularly fascinating.

In physiological terms, it's quite inexpensive to produce carbohydrates and pollen on poor soils, and so ericas are able to attract long-distance pollinators. About 75% of all bird-pollinated

The fire heath, Erica cerinthoides, *grows all over South Africa. One of the few ericas able to resprout after burning, its flowers are among the first to bring colour to the veld after a fire. The size and colouring of flowers, which may be white, pink or red, vary from region to region.*

plants occur in the Cape floral kingdom, where only two bird species are the pollinators of around 100 species of plant.

The most conspicuous pollinator is the Cape Sunbird, which specialises in proteas, but the Orange-breasted Sunbird (here the scientific name *Nectarinia* fits perfectly) is the one that visits the long-tubed ericas. These flowers have developed an abundant well of nectar concealed deep in their bright funnels. The deeper you must wriggle to reach the sweet reward, the better the odds that you'll emerge with a fair amount of pollen on your beak and the feathers of your face.

Then there are the hawk moths and bee flies, the colibris of the insect world. The species that visit ericas and other flowers have developed absurdly long mouthparts that fit exactly into the flower cavity.

There's also been some adaptation on the flower's side: groups of single-coloured flowers have evolved, usually pink or cream, which provide a sustainable source of nectar for the moths and flies. Like the birds, they're diurnal creatures and fly far and fast – the ideal vector in this finely balanced game of reproduction.

Of all our plant species, *Erica* has what probably seems like the most precarious way of life. An abundance of species crammed into a small and hostile habitat sounds like a recipe for disaster, but they have triumphed. Harnessing fungi at their roots, providing the sweetest of flowers at their tips, they have established insoluble partnerships with the world around them.

Simply irresistible.

THE HEATHER ON THE HILL

Heathlands occur all over the world, most natural, some created by centuries of clearing of woodland. Low-growing woody plants dominate, many of which are usually ericas, evergreen shrubs characterised by small, sticky, needle-like leaves, tightly packed on the stem. The upper surface of the leaf is waxy and curled slightly under to reduce evaporation. The bell-like flowers come in an astonishing variety of colours, shapes and sizes. In bloom they cover the landscape with soft washes of colour, most visibly lilac, purple and mauve, but the range includes nearly all colours, except for blue.

In Britain, and especially in Scotland, a rich nostalgia is associated with the heathlands as remote, wild places. There is also a keen interest in the plant life and heath gardens devoted almost entirely to ericas remain a feature of many large country estates.

Not so in South Africa, perhaps because the fynbos is the most diverse of all heathlands, so fynbos gardens usually tend to reflect this great variety rather than focusing on one particular family. There are, however, more than 600 species of erica native to Southern Africa, and for Western Cape gardeners in particular, the subtle pleasures of an *Erica* garden could be well worth exploring.

The inclusion of some natural companion plants such as restios, with their strong forms, plus drifts of watsonia and helichrysum could make for a striking picture through all the seasons of the year.

The mealie heath, Erica patersonii, *grows in the marshy coastal plains of the Western Cape. Natural habitat and populations have been severely reduced by urban expansion and farming and this beautiful erica is now listed as rare.*

ERICA BASICS

- In cultivation, ericas prefer rich, sandy, acid soil with a pH of around 5. The ground must be well drained. Avoid any spot that is inclined to remain damp.
- Prepare planting holes by working in plenty of acid compost and decomposed pine needles before you plant. Oak leaf mould will also work, if it is dug in well.
- An open, sunny spot, away from plants that have different requirements, is best. In regions with heavy frost, ericas will need a warm, sheltered spot.
- The *Ericaceae* have a shallow root system and the ground around the shrub should therefore not be disturbed.
- Avoid chemical fertilisers. A little bone meal in early spring and in autumn does help.
- These ericas are the easiest to grow in gardens: Berry heath (*E. baccans*); Bridal or Albertinia heath (*E. baueri*); Orange heath (*E. grandiflora*); Red hairy heath (*E. cerinthoides*); *E. vestita* and *E. versicolor.*

Clockwise from top left:
Erica sparrmanii *grows to about 0,5 metres and has greenish yellow flowers.*
Erica glandulosa *is a vigorous shrub that grows up to one metre tall.*
One of the most beautiful: Bridal or Albertinia heath, Erica baueri (= bauera).
Erica versicolor *is a strong grower and a lovely cut flower.*

WATSONIAS

The Fynbos Phoenix

Bulbs have the patience of Job. They're geophytes, so they are able to store essential nutrients under the ground, secure from most devouring bugs and beasts, while they wait for better days.

Better days mean time to bloom and, for watsonias, this is usually spring. In order to flower, especially in the fynbos areas where watsonias are most at home, there's one major requirement: light.

Undisturbed fynbos will burn every 12 to 15 years. These fires bring new life to the choked, overgrown veld because, with time, the canopies of taller plants will have grown together and this cuts off light to the ground and those species that grow near it.

After a fire in the fynbos, no plant grows back as quickly as the watsonia, flooding – as in Namaqualand – the entire landscape in a sea of colour.

The colours of flowers are virtually impossible to define. How, for instance, to describe the drift from pink to yellow on the petals of "Peace", my favourite rose? What would you call the blue of a strelitzia's crest petals? Azure? Reckitt's blue? Peacock blue? The colours of watsonias are especially tricky because, although there are only four basic colours (white, pink, red and salmon), light shines through the flowers' trumpets, making them glow with all sorts of nuances – and making the basic colour as ephemeral as an opal or a live coal under ash.

White watsonias are chalk-white and more: as white as the wing of a tick-bird, with a hint of pearl in their milkiness. It's a white like no other. Like the sound of silence. When you look at it long and hard enough, the nothing becomes something.

And the red? Pomegranate and tomato come closest but, once again, there's a hint of something else under the surface. Gold. A few other fynbos flowers, such as *Tritoniopsis* and *Homoglossum*, also have this hidden gold dust – if you know how to look for it.

To me, pink and salmon watsonias fall into this colour grouping too: in both there lurks a

Watsonia zeyheri *grows along the coast from the Cape Peninsula to Bredasdorp and Cape Infanta, usually in damp soil, in seepages or along streams. One of the shorter species (35-80 cm), it blooms from October to early December.*

hidden gleam, like the interior of a shell. Walk around the flower and see how the play of colour shifts and changes with the light.

So which colour would you plant as a symbol of your love? White? No. Because white reflects light and colour, making its heart black. Pink? No. Too Cape, too frivolous, too naughty – all cheap thrills and sweet champagne. Salmon? No. Salmon is of the flesh. It's too lush, too voluptuous. Plant red for your love, because names for reds whisper of seduction and enchantment: vermilion, scarlet, crimson, cerise … blood-red. Then, pray that your love will bloom and endure, but not like the watsonia, only after a fire.

Better still, forget about love and plant watsonias for no reason other than their wondrous colours. Wait for the early morning sun, sit low behind the plant, and take in the light through each flower.

LILY OF THE FIELDS

The watsonia is named after Sir William Watson (1715 to 1787), a doctor and botanist who first recorded this exquisite bulb in the Western Cape.

But watsonias are not only to be found in the fynbos. There are about 52 species indigenous to South Africa that grow from the Cape right up through KwaZulu-Natal to the escarpment of Mpumalanga. Some are evergreen, some deciduous; flowering ranges from spring right through to late autumn in warmer regions. Most species grow in mountain fynbos or open or rocky grassland.

Trumpet-shaped flowers on long stems are borne among clumps of sword-shaped leaves. Colours are usually shocking pink, bright orange and red, but there are also white, pale yellow and mauve species.

The larger species make wonderful garden subjects and are ideal for grassland and massed plantings. Plant them in drifts of one colour for best effect. There are also dwarf species that make beautiful pot subjects.

COLOURS OF THE RAINBOW

- *Watsonia aletroides* is red, grows 0,5 metres tall and blooms from September to November.
- *W. coccinea* is orange, grows to 0,4 metres tall and blooms from August to November.
- *W. densiflora* is a pink summer rainfall species that grows to over a metre tall and blooms for six months from approximately November to July.
- *W. hysterantha* (rooipypie) is red with a slim tube of a flower.
- *W. marginata* (fairy or broad-leaf watsonia) is mauve, grows to 0,8 metres and is a medium grower. It flowers only in October.
- When next you're relaxing or climbing on the slopes of Table Mountain, keep a lookout for *W. tabularis*, which means "of Table Mountain". The flower may be vermilion, scarlet or bright pink.

Watsonias are best planted in drifts of single colours, en masse or in a grassland garden. Watsonia fourcadei *is a tall, evergreen mountain species and a wonderful garden subject. It grows on the slopes of the Cedarberg and the Winterhoek, and flowers from November to January.*

WATSONIA BASICS

- The best time to plant watsonias is in autumn, once the soil has cooled down a little. Give the bulbs a sunny spot, good, friable, compost-rich soil; then sit back and wait for spring.
- All watsonias need to be well watered, particularly in their growing season.
- Like many bulbs, watsonias multiply and can be divided to make more plants – an economical way to add more colour to your garden.
- Deciduous species can be lifted and divided every two or three years, but evergreen species should be left undisturbed for five years.
- The foliage of deciduous species can be cut back after flowering.
- In his book, *Garden Plants and Flowers*, Keith Kirsten has excellent advice on how to keep watsonias in good flowering heart. Like many perennials, they are inclined not to flower when they've been transplanted. To prevent this, dig carefully around the root ball, then slice some of the outer corms off with a sharp spade – it's a wonder remedy.

Clockwise from top left:
Watsonia hysterantha *is a small species which flowers before coming into leaf.*
Watsonia borbonica subsp. ardernei *is a sought-after pure white.*
Watsonia wordsworthiana *is a tall, showy spring-flowering species.*
Watsonia pillansi *grows in the Drakensberg as well as the Southern Cape.*

Perfume and Panacea

On the day I turned 14 I got my first pair of glasses. They had heavy black frames like the ones Yves Saint Laurent wore. I took my spectacles out of their case, slid them onto my nose, looked out the kitchen door – and nearly fainted. I had to hang on to the door frame because, for the first time in my life, I could actually see the Langeberg.

Mountains had up to then been a blue-grey haze, now there was definition and texture. "Come, said the mountains, come closer." The mountains of the Boland are a storehouse of sweet-smelling savours: pelargonium, mint, wild rosemary – and buchu.

Boegoe, boechoe, boggoa, bookoo, bucchuu – look it up in Smith's *Common Names of South African Plants* and you will discover that buchu is by far the longest entry in this precious book. There are quotes from Kolbe (1707), who reported that San women dusted buchu powder over their heads to make them more beautiful, and from Engelbrecht (1926) comes the story how Koranna mothers would sprinkle buchu powder into the folds of their babies' skin to soothe irritation and chafing.

As a child I never encountered fresh buchu because I lived on the plain; mountain rambles were limited to summer visits to pools in shadowy kloofs and occasionally, in the autumn after a veld fire, to pick nerines. The only buchu that I knew was from my Ouma Lilly. A hypochondriac, every morning she would pour boiling water on dry buchu and drink it with a long face. It was good for everything, she said.

Then in my last year at school I was invited to pick bush tea in the Langeberg. On the way up I was stopped in my tracks by a cloud of intense fragrance. Your first encounter with buchu is a seminal one. From then on, a mere whiff is enough to pull you back again to a world that has its own wind and its own sky.

Outeniqua buchu, Agathosma ovata *"Outeniqua", covers the region with a fragrant sprinkling of white from June to September. An evergreen of upright growth habit, it will grow in sun or shade, and reaches a height of between one and two metres.*

My first backpack, a worn grey Swedish Bergans, still has buchu leaves in a pouch from when I went hiking in Kouga 30 years ago. I only have to rub those leaves between my fingers to bring that whole landscape sharply into focus again.

Buchu is our oldest bush remedy and possibly the original bosje from which the names bosjeman and later Bushman came. This perfume of the veld is a powerful panacea, balm and linctus. But the uses of buchu are not medicinal only. When you come across bok buchu (*Agathosma capensis*), pick a sprig or two. Take a farmyard chicken, tuck a lemon and the buchu into the cavity, roast it in a casserole and invite a few bush-loving friends for supper. Wait till they're all seated around the table, then lift the lid – and I guarantee you an evening of veld stories.

If testosterone had a scent it would be the smell of buchu. Buchu is reminiscent of tobacco, sweat, leather and crushed leaves. Classic "male fragrances" tend towards fern, grass and soap, and sweet-smelling classics like Old Spice. Buchu, while it remains unmistakably manly, borders on pungent. If rose and jasmine are the Victorian ladies of fragrance, and tuberose and tonka bean the modern Ms, buchu is undoubtedly Dirty Harry.

THE HOME APOTHECARY

Rutaceae is a botanical family rich in volatile oils secreted through glands on the leaves and fruits. Some of our most delicious perfumes – think of lime, lemon and orange – as well as powerful pungencies such as rue, belong to this family. As does buchu.

Derived from a Khoi word for aromatic herb, buchu embraces a number of fragrant small-leaved, small-flowered shrubs indigenous to the Western and Southern Cape. These comprise ten genera, but it is *Agathosma* (in particular *A. betulina* and *A. crenulata*) which is the true medicinal buchu and famed all-round cure. The Khoi chewed the leaves, the Dutch went one better and steeped them in brandy. Buchu has antiseptic and diuretic properties and remains a widely used home remedy, both externally as an antiseptic and internally for chest, digestive, kidney and urinary tract ailments, and general malaise.

BUCHU IN THE GARDEN

Buchu plants grow naturally among the fynbos, usually on mountain slopes. This makes them ideal for winter rainfall gardens, less so for other parts of the country, especially regions with alkaline soil or drainage problems.

The scent is there all year round, especially during damp weather. And then, every year from June to September, the plants explode into clouds of tiny white, pink or mauve flowers. Plant them along paths or near patios with scented-leaf pelargoniums and helichrysum, where they can be touched in passing.

Some coastal buchus are more adaptable to alkaline soil. These are *Acmadenia mundiana*, *Agathosma cerefolium*, *Euchaetis meridionalis* and *Coleonemas* (which are also among the best garden and pot subjects).

Steenbuck buchu, Agathosma capensis, *is a small rounded bush, usually less than a metre tall, with sweetly spicy leaves, which may be found throughout the Southern Cape, and even as far north as Nieuwoudtville. Flower colour may be white, pink, mauve or purple.*

BUCHU BASICS

- Plant buchu in direct sunlight, preferably in close, mixed plantings with other fynbos plants and shrubs.
- Well drained, well composted, slightly acid soil is a must for the majority of species.
- Pinch young plants back to encourage bushiness.
- Most fynbos plants hate root disturbance. Take care to avoid root damage when planting, and always keep a good layer of mulch on the soil.
- Pests are few because volatile oils discourage insect attack.
- Fynbos plants are susceptible to fungal attacks. Keep the soil cool with mulch and the leaves dry during watering.
- Buchus are easily propagated from seed or cuttings. Cuttings produce bigger plants more quickly, and are best taken in spring and early summer.
- Seed should be sown in autumn in a light mix of sand loam and compost. A smoke treatment will assist in germination, which takes between one and two months.

Top left and bottom left:
Agathosma ovata *"Kluitjieskraal" is a small, rounded shrublet with aromatic leaves. Clusters of small, starry pink flowers appear from midwinter to spring.*
Top right and bottom right:
The confetti bush, Coleonema pulchellum, *has feathery leaves and bears masses of tiny pink flowers from winter to spring.*

Smoke and Mirrors

I was just four years old when Aunt Ella first took me for a walk out of town, to the cemetery. One hand held mine fast; the other carried a shallow basket with flowers gathered from her Victorian garden: candytuft, asters and pompom dahlias. We crossed the Willem Nels River and made our way through the unmarked graves to her parents' headstone. After we had tidied the grave and arranged the flowers, we walked to the town common, which adjoined the graveyard. And it was here that Ella recited this rhyme to me:

Down among the rushes
A little dog lies there
His tail is quite still
His bones bare.

I burst into tears. In my fertile imagination, in every clump of reed and bush, a little dog lay dead. To this day, I can see it lying in the restios.

Restios bear male and female flowers on separate plants – though you would scarcely call them "flowers" as such. The family belongs to that highly specialised group of plants pollinated exclusively by the wind.

Both male and female inflorescences are brilliantly adapted: the male produces vast quantities of pollen and the female has no nectar-producing organs, only feathery styles made to sift pollen from the air. No moth, no butterfly, no bird visits these sombre flowers – each fibre is designed to exploit every whim of the wind.

Wonderfully, there is also another survival stratagem, a partnership with a humble insect. In many restios, the seed develops a fleshy base, the elaiosome, which contains oils that are irresistible to ants. As in the case of some other plants, as soon as the seed falls to the ground, these avid

Elegia capensis *grows in damp areas of mountain fynbos and is an attractive fast-growing species, reaching a height of up to two metres. It is one of a few restios that will grow in dappled shade as well as full sun.*

collectors carry it away to their underground storerooms. Here the elaiosome is devoured, but the rock-hard seed remains unused, until winter. Then, safe from the reach of seed eaters – and the devastating flames that ravage the fynbos – the restio germinates, and its survival is assured.

Restios are tough plants, made up mainly of stalk. To protect against drying winds, the leaves have disappeared and photosynthesis is carried out by the stems. There are animals – dassies and klipspringers – that eat the young growth, but there is also another, more secretive nibbler of the shoots. Walk into a patch of restios, fold back the foliage and you will see that the substratum is crisscrossed with neat paths. At every fork and turn lies a pile of stalks – the work of that gentle, popeyed fynbos rodent, the marsh rat. One wonders what special symbiosis might be at work here. Perhaps some of those hard seeds land among these piles and find fertile ground for germination …

When I think of restios, I think of the colour brown. Nature's palette is at its richest here. The first impression is of fresh rust, but as the wind blows, the swaying stems and seed heads flash terracotta, khaki, chocolate and amber.

Restios have been harvested in the wild for thatching and broom-making for centuries. Take a drive through Moedverloor and the bleak plains of Hopefield and you will still see the old reed houses everywhere. These thatching grasses were being used as shelter against the wind and rain long before European colonisation of the Cape. Walk into the cool of a house with a thatched roof and you will smell the veld. Sweep your yard with a real reed broom and you will discover that no man-made fibre can do the job as well – or with the same sensation.

FROM THE WILD TO THE GARDEN

Restionaceae is an ancient family of plants. Old as Gondwanaland, restios are one of the signposts by which scientists are able to map the formation and drift of continents. There are over 400 species, of which at least 320 occur in the Cape floral region. The rest can be found in Australia, Madagascar, Chile and Southeast Asia.

These reedy, tussock-forming plants range in height from 20 cm to over three metres tall, and their striking forms and textures make them equally suitable as single specimens or massed plantings. All species have male and female plants that may be bewilderingly different, making identification extremely tricky.

Designed for wind pollination, their inflorescences are not showy in themselves, but the long-lasting seed heads and gleaming papery bracts, which range from pearly white to russet and dark chocolate, offer a fascinatingly different kind of spectacle. You have only to visit the restio garden at Kirstenbosch to be seduced and inspired.

Restios defied cultivation until 1989, when it was discovered at Kirstenbosch that smoke was the trigger required for germination. This breakthrough came at the right time to catch a major new trend in horticulture – the growing popularity of grasses and similarly airy, mobile plants. Seven years after this breakthrough, the National Botanical Institute (NBI) was able to launch the family into the gardening world at the 1996 Chelsea Flower Show.

A sampling of four different restios (clockwise from top left: Hypodiscus, Staberoba, Cannemois *and* Elegia sp.*) reveals the variety of inflorescences and seed heads within the family. All species have male and female plants, the flowers of which can also look very different, making identification a challenge.*

They have since become all the rage, proving surprisingly adaptable, and there are now specialist nurseries in both Britain and America to meet the growing demand.

RESTIO BASICS

- In the wild, restios usually grow in poor, sandy soil, typical of fynbos regions, but they respond well to a slightly richer medium. While most species prefer a slightly acidic soil, pH seems to be less important than really good drainage.
- Plant them at the beginning of the rainy season (summer or winter rainfall) to give them a good start. Make the hole about 30 x 30 cm and mix the soil with about two spadefuls of compost. Keep the plant well mulched and avoid disturbing the roots.
- Regular watering will promote sturdy growth, but once established, restios are able to cope with periods of drought.
- Plants respond well to mild feeding via a liquid fertiliser such as Seagro or a slow-release granular fertiliser.
- All species are sun-loving, but there are a few that will take dappled shade: *Elegia capensis*, *Calopsis paniculata*, *Thamnocortus insignis* and *Ischyrolepis subverticillata*.
- Dead stems may be removed as necessary. Some species are self-cleansing, with new culms displacing the old ones.
- Many species make spectacular pot subjects, provided the pot is big enough. Experts at the National Botanical Institute recommend *Chondropetalum tectorum*, *Elegia stipularis* as well as *Thamnochortus cinereus*, among others.
- Restios may be propagated by division of the clumps or by seed, which should be sown in autumn, after soaking in the seed primer Kirstenbosch Instant Smoke Plus. Both are available from Kirstenbosch.

Chondropetalum tectorum *is one of the handsomest and most widely planted of the family; a fast grower that will produce spectacular fountains of stems within two years. Mature clumps will grow to a height of 1,5 metres with a spread of up to three metres.*

Coral and Sky Blue

Proteas might well have been designed by architect Norman Foster. But if I had to ascribe an architect to the disa, it would be Jorn Utzon. The median sepal is one of the most sublime shapes in nature, resembling at once a shell, a column and a sail.

The symmetry of orchids is only bilateral, like that of an animal. And if I had to ascribe an animal to the disa, I would choose a reptile – the fiery mouth of a dragon or the bright throat of a chameleon. Of all our hundreds of indigenous orchids, disas remain the loveliest and most compelling, from the tiny *tenella* to the giant *uniflora*, from the blue flame of *purpurascens* to the poison yellow of *aurata* – the shape, unmistakably disa.

It was in the Limietberg, next to a mountain stream, that I came upon my first *Disa uniflora*. I was out looking for peregrine falcons when my water supply ran out and I climbed down a kloof in search of a trickle. The water in the upper reaches of our rivers is as clear as gin; only lower down does it draw, as though from a tanning vat, a brandy colour from the humus. Blind with thirst, I crouched by the first pool with my flask and took a long drink. And then my surroundings swam into focus. There at the seepage, three disas grew against a mossy bank; six actually – because the water was like glass. That is how I like to remember them, disa red against moss green, floating above the still water.

The mystery of the disa is that such beauty and complexity can be produced in such impoverished soil. Our most spectacular disas grow in the Boland, where the ground is virtually sterile, where the entire landscape is regularly swept by fire and shrivelled by summer winds. But then again, if you look at the worldwide distribution of orchids, more than 22 000 species, it's no surprise that they are able to survive even here.

The disa has lived here for so long that an astonishing partnership has developed between it and an insect of the region. *Aëropetes tulbaghia*, the Mountain Pride butterfly, is the only insect

The Pride of Table Mountain was first collected by Michael Grubb in 1767 and sent to the great Swedish botanist Bergius, in Uppsala, who gave it the name Disa uniflora. *This species is the parent of numerous exquisite hybrids.*

pollinator of *Disa uniflora*. Pennington describes this butterfly as "enchantingly beautiful and one of the most magnificent *Satyridae* in the world". Obsessed with red, this beauty may also be seen visiting *Watsonia*, *Kniphofia*, *Crassula* and *Brunsvigia* species. It's a dream of mine to see, just once, the Mountain Pride perched on a red disa – a perfect porcelain motif for Charlotte Rhead and Crown Ducal.

Disas, like holidays, must be earned. Wait until January in the Boland and choose a shimmering hot day with no wind. Pack your rucksack with only the essentials – water, fruit, nuts perhaps. Choose your waterfall carefully; sift through the layers of your memory to find one where the sun will shine on the water at midday. Approach without looking up and climb until you reach the foot of the cascade. Then look up, and there, flaming through drops of crystal, you will see the disas: coral against sky blue.

DISAS AT HOME

To grow disas, an understanding of their habitat is vitally important. This is also another good reason to go and see them in the wild. Disas are terrestrial orchids with swollen rootstocks called tuberoids, a combination of root and stem tissue. They serve as underground storage organs, and are regenerated annually.

Of the 131 species in Southern Africa, cultivation and hybridisation is mainly confined to seven cross-fertile winter-growing species: *Disa uniflora*, *D. racemosa*, *D. tripetaloides*, *D. cardinalis*, *D. venosa*, *D. aurata* and *D. caulescens*. They are relatively easy to grow and are more or less evergreen.

Five of these species grow along streams and waterfalls in the winter rainfall area and are submerged in winter; the others grow in seepage areas with good drainage.

The roots are in sand, among ferns, restios and moss, so that the water is well aerated and filtered, cold and mildly acid. The high altitude ensures that there is always plenty of light and air.

Growth starts in late autumn with mature tuberoids producing leafy shoots. Replacement tuberoids also begin forming in winter. The growth rate accelerates in spring and flowering takes place from midsummer, with up to six flowers per stem. Stolons also emerge from the base of aerial shoots and from their tips, young plants develop. There are now numerous beautiful hybrids derived from these species in many different shades of red, pink, yellow and orange, available at orchid societies and specialist nurseries.

DISA BASICS

- Good air circulation is essential.
- Plenty of light: 50% shade cloth is the usual recommendation.
- Humidity: 50 to 70%.
- Pure water with no salts.
- Plastic pots are best and should be no larger than 15 cm in diameter for mature specimens.

Hybrids Disa *Robert Cywes and* Disa *Kewfoam in flower at the La Motte estate, which is now home to the collection of world-renowned grower Professor Sid Cywes. An active disa growing and breeding programme continues on the estate.*

- Use a growing medium that allows for adequate draining, does not rot when kept wet, and retains moisture. Most often used is a mix of two parts of an inert material such as quartzitic gravel or perlite, to one part of an absorbent, slightly acidic material such as sphagnum moss, milled bark or fern fibre.
- Keep moist at all times but do not water too much.
- Stand pots in a tray of damp sand for coolness and humidity.
- Feed regularly in the growing season with very dilute solutions of a liquid fertiliser such as Nitrosol, Multifeed 10 or Kelpak, or use a slow-release temperature-controlled fertiliser such as Horticote 7.
- Repotting should be done annually; divide plants in late summer and early autumn.

OTHER BEAUTIES

- There is also a whole spectrum of deciduous disas – winter and summer varieties – and they are found from the Cape to Limpopo, KwaZulu-Natal and beyond. Their flowers are generally smaller, little bejewelled columns that rise up out of the veld after rain or fire, solitary in places, elsewhere in their hundreds. Rare or common, they are all a delight to discover. Look for them in damp veld, marshland, in the vleis, seepages and mountain grasslands – many of these disas also like their feet wet.
- Both winter and summer rainfall species have a symbiotic relationship with mycorrhizal fungi in the soil that makes them extremely difficult to grow and sustain in cultivation. Expert growers have achieved some success with a number of the winter-growing species, but the summer rainfall species remain a challenge. In general they must be enjoyed where they belong, exquisite, enigmatic reminders of the complexity and fragility of the natural world.

Clockwise from top left:
Disa *Kirstenbosch Pride "Monterey".*
Disa *Unilangley.*
A large-flowered, pure yellow seedling from an unnamed cultivar at La Motte.
Disa *Elegant Beauty "Blue Blush".*

The Neon Queen

There can be no shadow without light, because without light there is only darkness. As soon as light appears, so does shadow – darkness clings obstinately to light. Open the curtains a chink, look at the patterns on your ceiling and suddenly you realise that shadows consist not of solid black but also of grey and white.

There is the umbra, black at its blackest, the penumbra, half-shade, and the antumbra or nimbus of light; examined, shadow loses its gloom. And it is this complexity of light within shade that makes the colours which happen to be there so much livelier. Because green stays green, few plants shine out. Here though, plectranthus is queen. In the shade her snowy cream, pink, mauve and lavender flowers have an electric luminosity, like neon.

Plectranthus make ideal garden plants because they grow easily from cuttings and require almost no attention. Nearly all species flourish in compost-rich shade. They have shallow roots and like regular water, but can also survive long periods of drought because they store water in their fleshy stems. They're also well worth planting for their attractive leaves, which range from dusty mint and cream to hematite and bottle green.

Because most plectranthus bloom only from late summer to autumn, plants with colourful leaves are a good choice because then you have interest almost all year round. Another characteristic of the leaves is their scent when you crush or brush against them, which is different for each variety. Of all their common names, I like only spur sage (muishondblaar). Fly bush and gossip plant are sadly lacking in imagination.

As if we haven't sung its praises enough, plectranthus is also fairly frost resistant. When winter begins to tighten its grip on your garden, these rewarding plants will still be in bloom. Only in the teeth of winter do they begin to look chilly and that's the right time to cut them back hard, to about one-third of their height. The ground-cover varieties need practically no care.

Bred by local horticulturist Roger Jacques, Plectranthus *"Mona Lavender" is a cross between* P. hilliardiae *and* P. saccatus. *This floriferous shrublet has proved a garden winner in both colour and growth habit.*

Plectranthus is also an ideal plant for propagation. Where a stem node touches the ground it will quickly put out roots, so striking cuttings is simplicity itself. Break off stems with two visible nodes and stick them into the ground without bothering about hormone powder. This is a plant for streets with an old-fashioned garden culture, where people will still come over with a prize flower, which is simply a way of getting to see your garden. Once inside, conversations begin which only another gardener will understand: "Mmm, yeees, I see …" Plant sheets of plectranthus so that you can have cuttings to give away. Gardening with indigenous plants is a noble calling because it demands vision and dreaming far into the future.

Plectranthus is one of our most precious indigenous plants. Precious because it is our own and because it is so tough and undemanding. But its greatest virtue lies in its ability to transform gloom into a play of light, to make a patch of shade shimmer like an opal – mauve and lavender, glowing like coral in the green depths.

SPUR OF THE MOMENT

Plectranthus (Latin for spur flower) belongs to the mint family, which goes some way towards explaining those fragrant leaves. In 1788 the French botanist L'Héritier, working from specimens of *Plectranthus fruticosus*, misleadingly coined the Latin name from the spur at the flower's base, a detail which most of the species do not have.

Of a total of 350 species, 53 are indigenous to South Africa. *Plectranthus* expert Ernst van Jaarsveld has coincidentally also declared these the most gardenworthy. They all grow in the summer rainfall regions, the majority in coastal subtropical forest, some in afro-temperate regions and a smaller number in bushveld or sunny grassland.

For gardeners, Ernst divides *Plectranthus* into four groups, according to their growth habit. There are shrubs and ground covers, both of which generally prefer shade, though there are exceptions in both groups. Then there are the succulents for rockeries, and another group that make good house plants.

Long neglected as garden plants because of their supposed susceptibility to frost, plectranthus have at last come into their own because of their toughness and their ability to grow in shade, which makes them ideal for small, sheltered city gardens as well as the heavy shade of our cities' man-made forests. A host of beautiful new cultivars have made them even more versatile and worthwhile.

PICK OF THE BUNCH

- *Plectranthus ambiguus* – A compact yet spreading forest-fringe shrub, with large mauve spires.
- *P. amboinicus* – Or Indian mint, a low-growing herb with succulent, deliciously aromatic leaves, for full sun or semi-shade.
- *P. ciliatus* – A low grower with white to mauve flowers and chestnut-backed leaves, from damp forest floors. It's a good ground cover for shady spots.

"Lilac Spur", a hybrid of Plectranthus oertendahlii, *is a spreading perennial with lilac pink flowers and the shimmering leaves of the parent species.*

- *P. ecklonii* – A superb shrub with mauve, pink or white flowers, for mass planting in partial shade. Prune after flowering to ensure vigorous new growth.
- *P. fruticosus* – A tall shrub with undersides of leaves tinted pink, and pink or blue-mauve flowers. Another one for mass planting, suitable for full shade.
- *P. lucidus* – A glossy ground cover ideal for coastal gardens.
- *P. madagascariensis* – A highly fragrant, velvety succulent ground cover for sun or shade.
- *P. neolichus* – A very useful low-growing grassland species, happy in sun or light shade.
- *P. oertendahlii* – Decorative leaves and a trailing habit make this a charming pot plant.
- *P. saccatus* – The popular stoep jacaranda has large flowers and delicate leaves, which make it a lovely pot plant, ground cover or edging plant.
- *P. verticillatus* – The gossip plant has fleshy round leaves and bears strings of white flowers with mauve flecks. A lovely ground cover where frost is mild.
- *P. zuluensis* (Zulu spur flower) – A showy shrub best suited to warm subtropical gardens.

PLECTRANTHUS BASICS

- Choose your species or cultivar to suit your climate and situation.
- Plant in large drifts and groups for maximum effect.
- The soil should be well drained and well composted with the addition of some kraal manure and bone meal.
- In the open ground, keep plants well watered until established. Then water only when the soil is dry.
- Plants in containers, however, should be kept damp.
- Prune back fairly hard in late winter.
- Pests and diseases are few but in certain soils eelworm may present a problem.

Clockwise from top left:
Plectranthus hilliardiae *grows in the deep shade of river gorges in KwaZulu-Natal and the Eastern Cape.*
Plectranthus ecklonii *is one of the largest and showiest species.*
Plectranthus neochilus *is a very useful and decorative ground cover.*
Dark green leaves offset the deep rose flowers of cultivar "Cape Angel Pink".

The Art of Evanescence

In Diego Rivera's magnificent painting "Mercado de Flores" the background is dominated by a dome of tightly packed arum lilies; in the middle ground a flower-seller smiles, her hair Latino black against the white trumpets. A young woman receives a sheaf of blooms, while a boy binds yet more stems in the foreground.

Look, however, at the melancholy girl on the left, the one with the giant bunch of tuberoses. Of all cut flowers, the arum is surely the most sensual and suggestive in form, to my mind much more so than the *Anthurium*. Despite this perfection of form, the arum has one fatal flaw: no fragrance.

Rivera, a renowned sensualist, must have known the scent of tuberose, which a perfume authority has described as "the most voluptuous of all flower notes. (…) It smells of very, very warm flesh". And so, in a marvellous moment of creation, he imbues both flowers with an even deeper allure.

Rivera was not the only artist enchanted by arums; Georgia O'Keeffe painted huge canvases with just a detail of the flower, incandescent and perturbing. This gift for reductionism was enhanced by her relationship with Alfred Stieglitz, the renowned photographer. Much later we have the photographs of Robert Maplethorpe: his studies of flowers, especially arums, are far more erotic than his more explicit images.

The arum is also known as calla, calla lily and pig lily, the name given to it by the early settlers, which persists in the common Afrikaans name, *varkoor.* I suspect that the pig referred to is the porcupine, which loves to dig up and devour roots and rhizomes. It's highly unlikely that the name refers to the domestic pig, because all parts of the plant are poisonous to both humans and most beasts.

And yet there are creatures which are largely dependent on arums. The oddest example is the arum frog, a tiny amphibian frequently seen inside the flower where it lies in wait for insects.

The common white Zantedeschia aethiopica *possesses a pure elegance that has made it one of the most popular garden plants and cut flowers in the world.*

The larvae of certain hawk moths are also partial to the leaves and a garden with arums would usually be filled with these lovely insects, tapping on your window on a summer's night.

The range of available species, cultivars and colours is now much more comprehensive than in the days when everyone got their arum bulbs from their neighbours. The yellow arum, *Zantedeschia elliottiana*, is much deeper and richer than mere yellow, and *Zantedeschia rehmannii*, my favourite, has flowers of glowing mauve that pales on the edges. Nowadays there are new cultivars so richly purple that they are almost velvet black.

It takes some doing to grow arums in the Karoo, but if you have a shady courtyard or a cool corner in the garden with a little pond, you will succeed. The effort would be worth it, because once a year you will have a living work of art, all the more valuable because it is ephemeral.

DECEPTIVE SIMPLICITY

What you see is neither a flower nor a lily, but a uniquely elegant botanical arrangement. The flowers are there, but minute, packed together on the central column or spadix and arranged in spiral zones of male and female flowers. The single "petal" furled around the spadix is actually an adapted leaf or spathe and provides a convenient cup for pollinating insects (usually crawlers), as well as a hunting ground for crab spiders such as our little frog (*Hyperolius hopstocki*).

This supremely stylish and successful floral arrangement belongs to the genus of *Zantedeschia* in the large family of the *Araceae*, hence the name arum. Zantedeschias are found only in Africa and they are by far the most appealing of this family of bog and jungle dwellers.

The differences among some of the *Zantedeschia* species tend to be subtle and botanists can't quite agree on how many there are – six, seven or eight. They are all deciduous with the exception of *Z. aethiopica*, which will remain evergreen if it gets enough water. This is the white arum which lights up vleis and streams of the Western Cape from about August, looking from the air like so many flocks of white birds. It also grows in damp spots throughout South Africa, which makes it a very adaptable garden plant.

There are also beautiful varieties of the arum lily which occur naturally; a pale pink one, which has been selected and registered as "Marshmallow" and a very large green and white form, which has given rise to the cultivar "Green Goddess". *Z. aethiopica* will vary in height (between 60 cm and a metre) and flowering time, depending on the place, but generally blooms from spring to midsummer.

The other six or seven deciduous species include *Zantedeschia pentlandii*, *Z. jucunda*, *Z. elliottiana*, *Z. albomaculata* and *Z. rehmannii*, all of which grow in the summer rainfall areas. They are smaller and daintier and range in colour from cream, yellow and greeny white to deep pink and plum. They may have plain or heavily spotted, highly decorative leaves, which are triangular in most species but narrow and lance-shaped in *Z. pentlandii* and *Z. rehmannii*. From these species, a range of dark and colourful hybrids, including some very sombre shades, have been produced by New Zealand and Dutch breeders. The dark colours are popular with florists but for garden and container value, the paler colours are the most effective.

"Pink Persuasion" is one of many sumptuously coloured new hybrids developed over the past 20 years.

ARUMS IN THE GARDEN

In large formal pots or lavish banks *Zantedeschia aethiopica* brings a wonderful cool lushness and opulence to a garden. This arum is at its most effective in an all-green planting, combined with ferns or asparagus species, with other white flowers, or with reeds and similar wetland plants on the rim of natural ponds.

Out in the wild, the summer rainfall species can often be found growing packed in the clefts of rocks, which shows how happy – and lovely – they would be as pot subjects. In the open garden they look their best in generous clumps and drifts.

All arums make wonderful, long-lasting cut flowers; the more you have to pick, the better.

ARUM BASICS

- Arums are rhizomatous plants and multiply freely.
- Coarse, well drained, well aerated soil is vital, or else rhizomes will rot.
- Deciduous arums must be kept dry in winter.
- *Zantedeschia aethiopica* – and its blush and green cultivars – needs shade for part of the day, articularly in summer rainfall areas.
- The coloured arum flowers best in full sunlight but should be given partial shade in hot areas.
- Arums are heavy feeders and need plenty of water, manure and compost.
- Give them an occasional fillip with 2:3:4 – this is a good balanced diet for bulbs, especially if they grow in pots.
- Arums may be cultivated fairly easily from fresh seed but division is the simplest route. Wait until the seed has matured before lifting and dividing clumps.

Clockwise from top left:
Zantedeschia *"Mango" is yellow flushed with orange.*
Z. albomaculata *is a summer rainfall species with creamy white flowers.*
Delicate pink Z. *"Marshmallow" is a form of* Z. aethiopica.
Z. *"Pot of Gold" is a cultivar with decorative, heavily spotted leaves.*

Crown Jewels of the Cape

Carl Linnaeus, virtuoso of names, extended his sublime artistry in classification of the natural world even to the elusive demesne of scent. In his *Odores Medicamentorum*, which appeared in 1752, he demonstrated his brilliant insight as an intuitive recorder of the ephemeral. His register of smell – Hircinus (goaty), Ambrosiacos (ambrosial), Testros (foul-smelling/stinking), Nauseasos (nauseating), Aromaticos (aromatic), Fragrantes (fragrant), and Alliaceros (garlicky) – remains spot-on even today.

Compare this alchemistic approach with the clinical entry in a modern textbook: "Olfactory information is processed in the central nervous system via pathways leading to the limbic system. It is not surprising that olfactory information should go directly to the limbic region, given that this is the part connected with the neuroendocrine regulation of both food-gathering and sexual behaviour." And after all that, they came up with only four smells: pungent, bitter, cool and sensitive. How very dull.

The world of perfume relies largely on plant and flower essences as well as skin secretions of animals. Classic scents like rose, jasmine and lavender have their origin in the perfume culture of Europe. Strange that flower perfumes from other parts of the world hardly feature, because if the perfumers of France were to encounter our wild irises, there would be some entirely new and wonderful fragrances on the market.

Our gladioli smell like nothing else on earth – sweet, but not cloying; strong, but not overpowering; and, above all, pure. Two stand head and shoulders above the others: the Marsh Afrikaner or aandblom (*Gladiolus tristis*) and the Large Brown Afrikaner (*Gladiolus liliaceus*).

Thirty years ago I went walking one summer evening on the mountainside above George. I remember the cool after the day's heat, the breeze on my face, the green of the ferns. And then I caught a breath of the most heavenly fragrance. Like a hound I went after it and there, among

In the fynbos, the smaller the plant, the more exquisite the flower: Moraea atropunctata, *no more than 20 cm high, flowers in August and September on clay slopes in the Western Cape.*

a clump of reeds, was the russet-speckled flower of the Large Brown Afrikaner. I picked it and carried it back to my room. After dark the sorcery began – the perfume grew stronger and the flower's colour changed from rust to underwater blue. At the stroke of midnight a hawk moth flew in at the open window, hanging like a hummingbird before the flower to insert its long tongue into the cup of petals …

These scents are not only to be found in our gladioli, but also in the moraeas. With these, the scent is more lingering – like the fragrance of old roses. Unlike gladioli, the blooms last barely a day and so are not suitable as cut flowers. But all these perfumed wild irises are worth planting: a pot of Afrikaners or Caledon bluebells on a windowsill against the light – crowns of chartreuse, goblets of sapphire.

Take a sunset stroll across the wide-open spaces of the Overberg, that rolling landscape with perfectly formed hills. Come to a reedy vlei and suddenly you'll catch an elusive, powdery scent – aandblom. Clearly a flower perfume, but cut with the metallic edge of citrus. As Luca Turin described Guerlain's Vetiver perfume: "At the same time distant and perfectly clear ..." And so it is with our *Iridaceae*. Once smelt, the experience goes deep, slumbers. One day, out of nowhere, a scent will trigger it to life and a complete moment from your past will leap up.

THREE-POINTED STARS

This large and diverse family of bulbous plants consists of at least 65 genera and 1 800 species distributed worldwide. No less than 46 of these genera are native to South Africa, with the greatest concentration in the winter rainfall region.

The complex colouring and amazing variety of the *Iridaceae* make them the crown jewels of the Cape floral kingdom. Some of these genera, like *Gladiolus* and *Watsonia*, extend to the summer rainfall regions, where there are also important (and often more easily grown) species to be found as well as summer-blooming genera like *Crocosmia*, *Dietes*, *Dierama* and *Schizostylis*, all of which are popular garden plants.

At first glance it seems impossible that flowers apparently so diverse should all belong to the same family. But a close look at their structure with a botanist's eye reveals the common threads.

The flowers have everything arranged in threes. Almost all of them have six sepals arranged in two rings of three. The two sets may be identical or markedly different, as in the *Moraeas*. They all have only three stamens. The leaves are long and narrow, parallel-veined and arranged in a fan in one or two rows. The seed capsule or ovary is set behind the flower and is divided into three chambers. Nearly all the *Iridaceae* grow from corms, which are modified stems; or rhizomes, which are modified runners.

Now for the differences.

Flowers may be borne in multiples on soaring spikes or they may be single beauties on delicate stems no more than 10 cm high. Large or small, the flowers are pretty showy, which is why they're so rewarding to grow. Larger species like *Dietes* and *Dierama* produce splendid fountains of leaves that provide texture and ground cover in the garden. *Iridaceae* that are widely avail-

---

Clockwise from top left:
The painted lady, Gladiolus carneus, *grows from the Bokkeveld to Riversdale.*
The bokmakierie, Witsenia maura, *is a rare, shrubby iris.*
Papery-cream Ixia flexuosa *may be found from the Cape Peninsula to Riversdale.*
Evening-scented Gladiolus tristis *grows from the Bokkeveld to Port Elizabeth.*

able as bulbs or plants – *Freesia*, *Sparaxis*, *Crocosmia* and *Tritonia* – are generally easy to grow. Rather more demanding are the smaller jewels of the winter rainfall area, which make beautiful pot subjects, and are generally only available from specialist growers.

WILD IRIS BASICS

- The secret to growing the *Iridaceae* – as with most plants – is to find out as much as you can about their natural habitat. They grow in widely diverse conditions that range from permanently moist to baking hot and dry. Make sure you do your homework on the microclimate – the soil, the aspect, the rainfall – and decide if you can replicate these conditions in a pot or in the garden before you go ahead.
- Prepare the soil and make sure the drainage is excellent.
- Ensure they stay dry in their dormant season and keep a close watch for pests and diseases.

The Caledon bluebell, Gladiolus bullatus, *grows from Kogelberg to Potberg and flowers from August to October.*

Royal Blues of Africa

"Plant blue and purple sparingly, O gardener." The African sun is so direct, so glaring, that blue tends to looks washed out, especially around midday. No wonder, then, that it is such a rare colour in the veld.

Gertrude Jekyll had misgivings about blue, even in the grey light of England and it's thought that Vita Sackville-West planted the Purple Border at Sissinghurst as a defiant response. She did say, though, that this border had to be seen on a cloudy day and that the visitor should look upon it as an old, faded tapestry.

Vita's unerring eye may be observed everywhere in this magnificent creation. The Purple Border was planted not against a flint wall but against the magenta of old brick, placing blue and purple in tension with the larger canvas of the garden, and thus firing the imagination. Vita achieves the same effect in the White Garden, where the visitor becomes snow-blind and imagines himself moving within a cloud. Climb the wooden stairs to her writing room in the medieval tower, and in each window you will see the coloured glass she placed there – the garden seen through filters, dream within dream.

We have two indigenous plants, agapanthus and plumbago, which represent the royal blues of Africa. The flowers of both are blue, but looked at closely, each is far more than simply a colour, just as the unique blue of any particular sky says many things. No wonder the blues of the ancient world were costly dyes reserved for royalty. Think of Phoenician purple, derived from the oxidized slime of a humble Mediterranean mollusc, *Murex brandaris*, and ultramarine from the ancient lapis lazuli mines of Afghanistan, the blue of Vermeer. Our own *Agapanthus inapertus* from the Graskop area is a velvet blue so dark, so intense, that it looks almost like lampblack.

We are so taken up with the bigger picture, the border, the avenue, that it can blind us to the detail. Pick a single agapanthus flower and look at it closely. The five slim petals radiate out in a

Inky-blue Agapanthus inapertus subsp. pendulus *grows in the Graskop region and blooms from midsummer. Long stems, which may reach as much as 1,5 metres, lift the flowers above the surrounding veld.*

sine curve, their edges slightly wavy. The loveliest thing is the play of blue within the petal, the central shaft of colour that washes out into the silver blue of the edges.

Look closely at plumbago. First, the strange calyx with its sticky, glandular hairs, so sticky that as a child my sister would tack them on her ears as earrings. The needle-thin throat opens into five rounded petals each, like agapanthus, with a dark seam in the centre. Stand back and let the flowers merge into one another like a cloud of heliotrope. We must experience colour like this more often to remember it, as a defined hue with shadings, but also as a primary colour.

Will you plant blue flowers in your garden for a moody Monday or *l'heure bleue*? Rather plant blue in great beds and sweeps; plant blue as an earthly celebration of our southern sky, that bluest of blues.

BLUE IN THE GARDEN

Flower colour and how we see it depends on the particular combination of saturation, texture, size and form, and how it reflects and absorbs light. Most flower blues contain some red, and so tend more or less towards purple but often defy definition within the spectrum. Pure blue itself (without any yellow or red) is relatively rare in flowers. Neither agapanthus nor plumbago is true, pure blue in this sense, but in the garden both are worth their weight in lapis lazuli, especially in the heat of summer.

AGAPANTHUS

Cometh the hour, cometh the flower … In the heat of midsummer, plantings in bloom seem like pools, rivers or patches of sky. The *Agapanthus* family is endemic to South Africa and consists of six species and a multitude of forms, from which numerous varieties and forms have been selected and bred.

Two evergreen species, the capricious *Agapanthus africanus* and the familiar *Agapanthus praecox*, are both from the Southern and Western Cape. Four deciduous species with starry or weeping flower heads, or a combination of both, occur in the grasslands and mountains of the summer rainfall regions: deep blue *A. caulescens* in KwaZulu-Natal, pale *A. dyeri* from the Mozambique border, delicate *A. campanulatus* in the Drakensberg, and *A. inapertus*, the lovely weeping agapanthus of the escarpment, which ranges from the palest lavender to inky blue.

AGAPANTHUS BASICS

- In the wild, agapanthus grows only where the rainfall is over 500 mm a year.
- In the garden, they need a deep drenching once a week in the growing season.
- Plant them in well drained soil, enriched with plenty of compost and mulch regularly.
- Full sun is best for good flowering, but there are cultivars of *A. praecox* that bloom well in semi-shade.

Clockwise from top left:
A whiter shade of royal: snowy Plumbago auriculata alba; Agapanthus africanus *"Albus".*
True blue Agapanthus inapertus *"Graskop", complete with sunbird;* Agapanthus praecox.

- Garden-collected seed is often not true to type. The only way to be sure that new plants will be the same as the parent is by division.
- Lift and divide evergreen types every four years, in late summer after flowering. Deciduous types should be left undisturbed for longer – six or even eight years – before being lifted and divided in spring.
- Cut deciduous agapanthuses back at the end of the growing season and cover with a thick mulch of compost.
- Most species make superb container plants.

PLUMBAGO

Plumbago auriculata is the star of a genus of ten species, a scrambling shrub indigenous to the valley bushveld thickets of the Eastern Cape and also scattered elsewhere in the summer rainfall regions. It combines resilience with an exuberant, cloudy delicacy that has made it a popular garden plant all over the world.

Extremely versatile, plumbago can be allowed to climb into trees, to ramble or sprawl across banks (it spreads by suckers) or clipped into a formal hedge or an informal billow. It responds extremely well to clipping, which in summer will always bring on a fresh flush of new growth and flowers.

The pale trusses of flowers have an extraordinary lightness and coolness about them. This is one case where washed-out blue works beautifully because of the shape and distribution of the flowers on the plant. Give them a dark-green backdrop if you can and don't be tempted to go for the lurid, darker blues; they do not work nearly as well, if at all, in a garden setting.

PLUMBAGO BASICS

- Plumbago responds superbly to heat, looking cool and fresh and flowery even in a Karoo midsummer.
- Somewhat frost-tender, it's not as keen on cold and needs cutting back in late winter or spring, when it will regrow quickly.
- Plumbago is propagated easily from seed, cuttings or suckers. The easiest way is simply to remove a rooted sucker from the mother plant in early spring.
- Plant in well drained, well composted soil.
- Keep fairly dry in winter but well watered in summer.

The airy simplicity, versatility and resilience of Plumbago auriculata *have made it one of the world's most popular garden plants.*

Birds of Paradise

If you have a small garden, don't even think about planting a strelitzia, especially *Strelitzia nicolai*. As a chick in the nursery, this huge bird will look completely innocent; so small that you can tuck it into the boot alongside the petunias. Mistake number two is when you plant it in a corner of the garden, right next to the wall. Three years later you have a green monster towering over the garden and a wall in ruins. But what a monster …

Also known as the giant bird of paradise, or wild banana, in a big garden this plant is one of the most effective ways of creating a tropical mood. The whole plant is pure architecture – the ribbed fans of giant leaves look exactly like monochrome props for the Folies Bergère. Coucals hide among these fans; waxy purple and white, the flowers ooze nectar which sunbirds come to drink.

One of the five species of strelitzia indigenous to South Africa is surely one of the best known plants on earth. Together with roses, irises, orchids and Heliconia, *Strelitzia reginae* is ranked internationally as one of most desirable of cut flowers – and with good reason. It is cultivated extensively in California and Hawaii and shipped to the cities of North America. I can't think of one issue of *Architectural Digest* without a crane flower in it. Any Art Deco flat looking out over Central Park will also have them, in a Steuben vase against a wall of glass.

Even without flowers, these plants are fascinating. The leaves and stems ask to be touched, every ridge and seam to be fingered. But it's the flowers that are the most compelling. In the whole plant kingdom there is no other inflorescence so complex or abstract: bird, fighter plane, dragonet? No, the flowers make me think of those fabulous luxury cars of the 1930s: Cord, Duesenberg, Auburn, the Bugatti Atlantique. At the head of every bonnet there would be a winged ornament in chrome, sometimes with insets of Lalique. The theme was flight, speed and elegance, with names like Goddess of Speed.

The horizontal inflorescence, plum-coloured, with a fine layer of wax, serves as a capsule

With its narrow, lance-like leaves, Strelitzia juncea *is the most elegant of all the species. In the wild it is restricted to a small area near Uitenhage.*

and platform for the orange and blue petals (the blue is like an electric spark, the orange neon). Sunbirds, the plant's only pollinators, perch on this magic box of tricks. First, the little winged jewels are drawn by a lavish supply of nectar and then, as they feed, concealed triggers release the pollen: a seamless partnership between two of Africa's most exquisite organisms.

And the name strelitzia? The hapless plant is named after the Duchy of Mecklenburg-Strelitz, the birthplace of Queen Charlotte of England. Yet another example of botanical colonialism? No, because the name is worthy of the crane flower; Charlotte was one of the founder members of Kew Gardens and also took music lessons with Johann Sebastian Bach. Strelitzias are timeless classics, like a handmade Italian shoe, or the mystery of Dior's Dune late at night, after a dinner party.

STRELITZIAS IN THE GARDEN

Strelitzias bring a spectacular architecture to the garden scene. Their unique flowers are remarkable not only for their form and coloration but also for their botanical engineering. Take care to plant them in the right place and give them plenty of elbowroom. Mature clumps of these perennials may reach three metres across, with fleshy roots extending several metres in all directions. Moving them, even dividing them, once the clumps have reached a good size, is virtually impossible without heavy machinery.

In her garden in Kimberley, Strilli Oppenheimer has devoted a walled area entirely to strelitzias, in a simple yet magical planting. All the species are here, among various species of wild asparagus, particularly *A. densiflorus*, a fine foil for their strong forms. *Bulbine frutescens* is the dominant ground cover and creamy green columns of *Eucomis autumnalis* appear from midsummer.

The Chinese plumbago *Ceratostigma willmottianum* will also provide an excellent foil for *Strelitzia reginae* (see below). Its bright blue flowers and pinky coppery foliage tints echo exactly the crane flower's colours.

BIRDS OF A FEATHER

Strelitzias belong to the large-leaved, largely subtropical banana family, *Musaceae*. There are five species indigenous to South Africa, two of which are stemless perennials and three, trees which sucker from the base. The differences between these latter tall-growing species are quite subtle, however.

- *Strelitzia reginae* is the most familiar, a stemless perennial that forms large clumps of grey-green leaves, the central ribs of which are decoratively pink when young. It grows in grassland, bush and thicket all along the east coast. Flowering is best in full sun, but it should be planted in semi-shade in hot dry areas. The characteristic boat-shaped bracts are plum and green, and produce multiple orange and blue flowers in succession. The *Strelitzia reginae* called "Mandela's Gold" has yellow rather than orange petals.

Each boat-shaped spathe of the crane flower contains between four and six flowers which emerge one at a time. Each flower consists of orange sepals and blue petals, and contains a gland which produces copious amounts of nectar, a magnet for birds, which serve as pollinators.

- *Strelitzia juncea* has similar flowers but narrow reed-like leaves and is more drought resistant. In the wild it is restricted to a small area near Uitenhage and Patensie. In regions of higher rainfall the soil must be very well drained. Plant in full sun.
- *Strelitzia nicolai* (the Natal wild banana) is common in coastal dune forest from East London northwards and grows up to ten metres tall, clumping from the base. It has spectacular fans of broad leaves, blades that are two metres long. These "tear" in the wind to create a palm-like effect. The bracts are dusky plum, the flowers pale blue and white. One spathe grows up out of another, creating a multistorey effect. It will tolerate drought but not frost.
- *Strelitzia alba* is similar but smaller, growing only to six metres in height. Its flowers are simple, not compound, and the petals are all white. In the wild it is confined to a small area between George and Humansdorp.
- *Strelitzia caudata* is the rarest of these species, found in the mountains of Mpumalanga and Swaziland. It grows to eight metres tall, also branching only from the base. Bracts are brownish red and the flowers, pale mauve and white.

STRELITZIA BASICS

- Well composted, well drained soil.
- Plenty of room.
- Full sun for good flowering but semi-shade in really hot, dry areas to reduce leaf burn.
- Do not disturb – plants are relatively slow growing, resent transplantation or movement (if you can move them) and take at least three years to flower from planting.

A litany of leaves (clockwise from top left):
Strelitzia nicolai; Strelitzia reginae; Strelitzia juncea; Strelitzia reginae var. parvifolia.

The Cloth of Gold

If Africa's trees are her cathedrals, then her vestment is the grass. As with any cloth, it is at its most beautiful sun-dried and rippling in the wind. In the Kaokoveld, I once woke at sunrise in the turning of the year. Propped on my elbow I lay and looked at the world; it had been a summer of good rains and the plains were covered with golden grass. The wind began to blow lightly from the east – the first ripple of the day. Gold plate before the wave, gold dust behind.

Grasslands were humanity's first breadbasket. Was it in Mesopotamia that we first planted wild grass around our separate dwellings? Or did seed fall from the thatch and germinate along the drip line? Fertilised by the dung of our first livestock, were the grains larger and fatter? Here, grass and stone and wind came together; and we were able to use the wind to winnow and the stone to grind bread from the earth.

Gilgamesh, mad with grief at the death of Enkidu, calls to Siduri: "Why should I not roam the wild pastures in search of the wind?" Grass, generous, fertile, ephemeral, was already a central theme of the imagination.

Grasses are among the most primeval forms of life on earth, the most widely distributed, and so old that they were grazed by the first dinosaurs. So old that people began to cultivate grains 10 000 years ago; so old that grass stalks were possibly used to make the first paper. The Bible is full of references to grass and its transience, brief as the life of man. In the cemetery of my home town, the words of the Psalmist are inscribed on the worn gravestone of a child:

The days of man are but as grass;
For as soon as the wind goeth over it, it is gone.

Our imaginative world is filled with grassy themes and references. Think of the heavenly bamboo-brush drawings of classic Japanese art, or the starlit cornfields of Vincent van Gogh.

Wild oats, Avena fatua, *is one of the many grasses that have colonised the world. Closely related to cultivated oats,* Avena sativa, *it grows in disturbed ground, particularly in sandy soil.*

Millais's jewel-like paintings of dreamy Victorian maidens define the period, especially the drowsy outdoor scenes of young girls reclining in long grass; high erotica in coy disguise. It's 30 years since I first saw Co Westerik's series of paintings "Snyden aan gras" (Cutting on grass), but they still make me shudder. Look hard at Ina van Zyl's "Pramberg" and the canvas begins to ripple beneath your gaze.

When I dream of Africa I think of thatching grass, bristle grass, old man's beard. I imagine myself walking through the Kilombero Valley where singing women harvest rice in the muddy fields; in Equatoria where violet-eared waxbills flit like sparks through the grass stalks; along a gin-clear lake in Uganda, finches in the undergrowth.

Iphigenia walks to meet Agamemnon, on her way to be sacrificed by the high priest of the oracle. Agamemnon takes her bridal wreath of dry grass and leaves, places it on her lovely, curly head and sprinkles it with holy water. Turning to the sun, she says, "Farewell my beloved light," and walks the last few steps up the mountain. Cachas waits in a cloud of smoke, a knife under his robe. Iphigenia enters the smoke and screams once. And the wind begins to blow. Our bond with the grass and the wind is old indeed.

THE VELD IN YOUR GARDEN

Grasses have huge value in the garden. Elegant, airy and resilient, with a long season of growth, they bring a touch of the veld to city gardens and, in an increasingly deforested urban environment, they also provide a habitat and resource for our numerous grassland birds and insects.

The ornamental use of grasses in landscaping was pioneered in the 1950s by German nurseryman Karl Foerster, who combined them with perennials in naturalistic plantings. In the US, Wolfgang Oehme and Karl van Sweden, inspired by the American prairies, planted grasses in blocks and large drifts with prairie annuals to spectacular effect.

Since then, in the hands of designers such as Piet Oudolf, Tom Stuart Smith and South African Patrick Watson (who designed the grassland garden at the Aparthcid Museum in Johannesburg, for example), grasses have transformed the face of landscape gardening. There's also a trend towards recreating the natural veld, as in the remarkable habitat created by John Masson of Ecoscapes at the Royal Canin Eco Industrial Park, North Riding, which won first prize in the 2007 Pam Golding Gardens of Pride competition.

Whether you choose to establish a grassland in your garden, or simply to use indigenous grasses among other plants, you are set for a whole new world of interest and pleasure. You need to observe and learn as much as you can about the natural habitat of plants and work with the local ecology in your design, advises expert Dawid Klopper, head gardener at Brenthurst (where the bold plantings of Highveld grasses in 2002 rocked the Johannesburg gardening establishment).

"It is good to work out a strong idea of what you would like to create, but a grass garden is ultimately a process rather than an end result," he explains. "Nature is dynamic and you should simply seek to steer the natural process of succession."

Clockwise from top left:
Weeping love grass, Eragrostis curvula, *is a key grazing grass.*
Ripgut brome, Bromus diandrus, *is an exotic grass that has become a weed.*
Root systems of perennial grasses may extend two metres deep and wide.
Field of inspiration: Avena fatua *and shorter* Briza maxima *sway and sparkle.*

INDIGENOUS GRASSES TO PLANT

These are Dawid Klopper's choices:

- *Ehrharta erecta* (Lamarck's Ehrharta)*
- *Eragrostis chloromelas* (Curly leaf)
- *Eragrostis racemosa* (Narrow-heart love grass)
- *Eragrostis superba* (Saw-tooth love grass)
- *Hyparrhenia hirta* (Thatch grass)
- *Hyparrhenia tamba* (Blue thatch grass)
- *Melinis nerviglumis* (Mountain redtop)
- *Melinis repens* (Natal redtop)
- *Miscanthus capensis* (East coast broom grass)
- *Panicum maximum* (Guinea grass)*
- *Setaria megaphylla* (Ribbon bristle grass)*
- *Setaria sphacelata* (Golden setaria)
- *Sorghum bicolor* (Wild sorghum)
- *Themeda triandra* (Red grass)

* Suitable for shade

GRASS BASICS

- Most grasses need plenty of sunlight and long hours of daylight. There are a few species that tolerate shade, so choose carefully.
- Indigenous grasses are usually difficult to grow from seed. It's often best to buy plants and allow them to drop seed, which somehow works. If you have grasses on site, make every effort to retain them, even if you have to lift and bag them.
- Too much compost and too much water will cause grasses to grow too quickly. They will then flop and have to be cut back. Dig in plenty of sand when planting grasses in established flowerbeds and do not overwater.
- If the new season's growth is still very lush, cut it back (as would have happened in the veld with grazing) to strengthen the plant.
- Long grass can be cut in late autumn, but always leave some clumps as a refuge for birds, insects and small mammals. Cut these in early spring.
- Add grassland bulbs, annuals and perennials to complete the picture.

Grasses differ from most other flowering plants in that they are wind pollinated. Their inflorescences are inconspicuous but exquisitely complex and mobile to facilitate the dispersal of pollen in the slightest breeze.

Barbed Beauty

The Old Testament bristles with references to thorns. You have only to look at Cruden's great *Complete Concordance to the Holy Bible*, in particular Isaiah 34:13: "And thorns shall come up in her palaces, nettles and brambles in the fortresses thereof: and it shall be a habitation of dragons, and a court for owls."

Thorns and thistles were a curse more fearful even than death or the plague. Thorns infested the humble patch of the first farmer, choked the crops, ruined the pastures. For the nomads of long ago, spring was a fleeting abundance; as the summer briars took hold, they had to move on, or flee.

How and where did thorns come into being? Primitive plants all had their origins in the algae. But somewhere in the past some of them became more palatable, and so more sought-after. And this is where the arms race began, when incidental thorns gave certain plants a better chance of survival. Plant eaters had to adapt, developing small mouths to graze between the barbs and robust teeth to crack them. Thanks to this ecological tug of war, today we have some of the most beautiful plants on earth: the acacias, the heavenly euphorbias of our deserts, the silk thistle and the rose.

Consider, also, Africa's classic thorn trees. You have only to look at the silhouettes. There is nothing accidental about these divine shapes, because they have been carved out by others in search of the tree's hidden treasure. Under the goad of the spines the lower layer has been pruned away and the crown compelled upwards. The whole landscape is defined by the finer detail, rather than by the big picture – the sky blue as nowhere else on earth, the forms pure Pierneef.

If I had to choose my favourite thornbush it would be the num-num. Was there ever a more rewarding and generous garden plant? You wake one summer's morning with the bedroom

No mystery as to why Euphorbia milii var. splendens *is popularly known as crown of thorns. Stems bristling with vicious spines are bare in winter, but produce bright green leaves and brilliant flowers like drops of blood come spring.*

windows open to the garden. The curtains billow and suddenly you're enveloped in a cloud of the purest perfume. It reminds me of Guerlain's Vetiver, as described by Luca Turin: "Stoic and discreet, Vetiver scorns all luxury save that of its own proud solitude". Later in the season, you are rewarded yet again with the ripening of the num-num's startling red fruit. Sweet and sour hidden among the double thorns.

We tend to forget that our world is a thoroughly prickly one. From the rose on your coffee table, the berries in your yoghurt – blackberry and raspberry – and the sweet chartreuse-green prickly pear, to the landscapes of Africa passing your car window, from fynbos to the Kalahari, thorns are everywhere.

A prick of the finger reminds us of our frailty. First the electric sting, then weakness – then the drop of blood, round and perfect. Redder than coral, num-num red. There is a beautiful verse, "Lascia la spina", from Handel's oratorio *Il Trionfo del Tempo e del Disinganno*:

Leave the thorn and pluck the rose,
You go in search of grief.
Old age will creep up on you
When your heart does not expect it.
(Bernadetto Pamphili)

NATURE'S DEFENCE MINISTRY

In September 1908, Sir Francis Darwin, botanist and son of the great Charles, made a startling claim. In an address delivered to the British Association for the Advancement of Science, he suggested that plants were able to "think" and that, as regards relation to environment, they and humans must therefore be placed in the same class.

This provoked howls of scorn. Take a look, however, at the defence strategies of plants and you might be inclined to agree. This armoury must count as one of the most extraordinary feats of evolution. Shoots, leaf stems and leaves have been changed into sharp protuberances, some hair fine, others large as lions' teeth; some straight, some wickedly hooked, others forked.

Over time, a live, soft appendage has become a hard, fibrous one with fortified cell walls, cunningly arranged to repel or ensnare – or both.

BASIC BARBS

Botanists refer to these modifications of plant parts as spines, specifying whether they are leaf, shoot, petiolar (leaf stalk) or stipular (a tag at the base of the leaf stalk). There are also thorns that are not modifications of standard plant parts but extra growths, termed emergences or prickles. Rose thorns and surface thorns of aloes are typical emergences. A spine generally contains some vascular tissue; an emergence does not. Mature spines and emergences are usually made up almost entirely of dead cells.

The leaf margins of Mexican desert dweller Agave parryi *have developed sharp teeth (these are emergences) to protect their succulent leaves from being devoured.*

Most of these elaborate defence mechanisms have developed in arid regions, along with other modifications designed to cope with drought. Stems swell to become water reservoirs, leaves become thorns, reducing transpiration as well as protecting the plant from thirsty animals. Other plants, such as acacias, retain their leaves, but the stipules change into large protective spines at the base. The spacing and angling may not entirely deter browsers, but will certainly inhibit them. Other thorns, such as the diabolical devil's claw (*Harpagophyton procumbens*) and the dubbeltjie (*Tribulus terrestris*) are designed to catch onto an unfortunate passer-by and so effect seed dispersal.

Hooked thorns along stems, such as in roses and asparagus species, are not only protective but also climbing aids. For all the animals that are discouraged by the barriers of the plant world, there are probably as many that use them as protection. Observe the number of birds nesting in thorny scrub and acacia trees: firm anchor pins to hold a nest – and safe as houses.

ARMOURED GARDEN BASICS

- In the garden, sculptural plants such as aloes, cacti, thorn trees and agaves provide dramatic beauty as well as security fencing, but keep them away from paths, patios and play areas.
- Keep branches of trees well above eye level, or let them sweep to the ground.
- Arm yourself with heavy leather gauntlets for gardening.
- *Aloe arborescens*, the cat thorn, *Scutia myrtina*, the Kei apple, *Dovyalis caffra*, or various taller species of *Euphorbia*, all make striking and impenetrable hedges.
- If you have space, plant a thicket or exclusion area of thorny trees, shrubs and scramblers, which birds will love.
- Begin by planting thorn trees close together along the boundary. To the front, plant shrubs such as *Cassinopsis ilicifolia* and *Carissa macrocarpa*. Add thorny scramblers, such as *Asparagus falcatus* and *Smilax kraussiana*.
- Spiky exotics also make colourful security hedges and habitats, especially bougainvillea and heritage roses. Gwen Fagan suggests planting a mix of the following old roses to produce a beautiful display: Géant des Batailles, Albéric Barbier, Indica Major, Mermaid, *Rosa laevigata*, *Rosa multiflora* varieties and Russelliana.

Clockwise from top left:
The bare stems of the crown of thorns, Euphorbia milii var. splendens.
The thorns of Acacia karroo *are modified stipules and may be as long as 100 mm on young plants.*
The sculptural naboom, Euphorbia ingens, *has diverging spines along the ridges of its swollen branches.*
The forked spines of the num-num, Carissa macrocarpa, *are modified shoots.*

The Torch Bearers

Every morning when she wakes, Baby Suggs, Tony Morrison's sublime character in *Beloved*, decides on a colour. Then she lies the whole day in her cheerless room and thinks only of orange, or yellow, or blue. So mauled and trampled by life is she, that only the world of her imagination can give her comfort. In my mind's eye I see her room: iron bedstead, ragged rug, frayed curtains and, floating against the ceiling like ectoplasm, a panel of colour, a vast Rothko.

When I call aloe colours to mind, I think of red, yellow and green. All these colours have something in common. They are intense, indelible and veiled in a layer of wax. Aloes are succulents and therefore resistant to dehydration: thick, fleshy leaves covered in thorns, stems like the tail of a rock lizard, scaled with the vestiges of dead leaves, delicate flowers, each with a waxy bloom. Run your hand over a spike of flowers and listen to the silky rustle, just like a chincherinchee.

Aloes are plants of the wild, Africa's cacti. Think of the Richtersveld and it is the aloes that define the landscape. You recall not forests of kokerboom, but a single tree against the sky. The tall trunk with its coffee-coloured scales, the characteristic cupola of sludgy green leaves. During a cycle of good rain they bloom abundantly – chrome-yellow flowers against the desert-blue sky.

Climb the highest peaks of the Boland, and there, along the crests, you will find aloes. Not classic rosettes of thorn-covered leaves around a crinkly trunk, but smooth leaves arranged in fans, with a fine-toothed edging. Old plants develop monumental shapes, mandalas against the rocks, fan upon fan. *Aloe plicatilis* is the most undervalued of pot plants, pure abstraction and the perfect form for a modern setting.

There is one region, however, where aloes not only define the landscape but dominate it. Late winter in the Southern Cape produces one of the most spectacular floral displays imagin-

The Wylliespoort aloe, Aloe angelica, *lights up the Soutpansberg in June, bearing a glowing candelabra of bicoloured flower heads above its crown of elegantly recurved leaves.*

able. *Aloe ferox* forms forests here, plants with tall trunks burnt clean by recurring fires, bearing above the rosette of leaves, a fiery candelabrum of flowers.

Beyond Caledon, in the region known as the Ruggens, with its lovely convoluted horizons, there are great ranks of them. When next you drive that way, stop next to one of these banks of flowers, and wait. Soon you will hear a sharp metallic call. Stay dead still and the Malachite Sunbird will come to perch on the stalks and tap his curved beak into the flowers. Nowhere else on earth will you see such a green against such a red against such a blue.

A little while ago a slip of paper fell from an old diary. I picked it up and sniffed it half in thought – a fleeting scent of vanilla. Guerlain's Shalimar! Years ago I had smelt it in Santiago and now another world, another moment, returned to me complete: an evening in Bellavista, a room in a dusty museum, the sleeping city at the foot of the Andes.

In the same way, an aloe flower would take me back to Weenen in the Gamsberg, to a lichen-covered rock in the Baviaanskloof or to the Overberg, with its grey-wing francolins calling at break of day.

BEAUTY IN ADVERSITY

Aloes are great survivors, equipped with an armoury that places them among the most evolved of plants. Their defence and recovery mechanisms include thorns, succulent leaves and a bitter sap that also contains a powerful healing agent for damaged and bruised leaves. This juice has been used medicinally for centuries and may be applied fresh from the leaf to cuts, burns and insect stings.

This most characteristic of African plants is also able to produce spectacular flowers in the harshest of environments. Most species seem to bloom when their nectar is most needed by birds and insects; in winter in the summer rainfall regions and in midsummer in the parched northern reaches of the winter rainfall area.

There are more than 550 species of aloe, most of which grow in Africa, plus a number in the Arabian Peninsula and Madagascar. About 125 species are native to Southern Africa. Aloes have assumed many different forms and growth habits in adapting to an amazing variety of arid conditions. Botanists have helpfully divided them into ten groups, but this doesn't always make for easier identification of individual species, especially among the speckled and spotted aloes. Aloes tend to be promiscuous and species cross naturally with one another, which can also make identification tricky.

ALOE TYPES

- Tree aloe – Five spectacular species which shed their dead leaves, highlighting their sculptural branch patterns.
- Single-stemmed aloe – About 15 species with tall, upright stems clothed in dead leaves and a single large rosette of leaves.

One of the few aloes found in the fynbos, the uniquely sculptural fan aloe, Aloe plicatilis, *blooms from August to October and is endemic to the mountains of the Western Cape.*

- Multi-stemmed aloe – Nine species which branch at or near ground level and grow into medium-sized or large shrubs or low stands.
- Rambling aloe – Five species with thin stems which scramble up through other plants or twine themselves into large clumps.
- Creeping aloe – Seven species with stems which creep along the ground but have upright blooms.
- Stemless aloe – About 25 species with rosettes on or close to the ground.
- Speckled aloe – Five species with short stems and spots or speckles on the surface or underside of the leaves.
- Spotted aloe – No fewer than 26 species, usually stemless, with heavily spotted leaves.
- Dwarf aloe – Six species in South Africa but many more beauties from Madagascar as well as Ethiopia and Tanzania make them a wonderful subject for collection.
- Grass aloe – A total of 24 mostly high-altitude species that flower in summer and are usually found in grassland.

ALOES IN THE GARDEN

Aloes are extremely easy to grow and effective in many different garden situations. They make superb accent plants and pot subjects. Groups of different species of different heights and forms look wonderful planted together. Spreading aloes, that is those which proliferate from the base, are excellent as ground covers and soil stabilisers. Single and multi-stemmed aloes are effective barrier plants and make attractive and impenetrable hedges.

ALOE BASICS

- Nearly all aloes require sun for all or part of the day to grow and flower, although some of the ramblers will grow in light shade.
- Choose species indigenous or suited to your area, taking frost and rainfall into account.
- Plant in good, well drained soil and mulch regularly with mature compost.
- Aloes may be easily propagated from seed, offsets or stem cuttings.
- Allow the base of offsets and cuttings to dry out before planting. Rather than bury the stem, simply place it on the ground, supporting the rosette with rocks if necessary.
- Aloes are able to withstand long periods of drought but with regular watering will grow better and be more pest-resistant. Allow the soil to dry out between waterings.
- White scale may be removed with non-toxic oleum.
- The most dangerous pest is the snout beetle, which lays its eggs in the crown. To save the plant, systemic insecticide must be applied via holes drilled in the stem.

Clockwise from top left:
Aloe microstigma *has densely spotted leaves and grows from the Western to the Eastern Cape.*
Neat and wickedly thorny, Aloe melanacantha *is a native of Namaqualand.*
In winter the giant Aloe ferox *blazes all the way from the Southern Cape to KwaZulu-Natal. (Bottom left and right.)*

Basilicas of the Desert

Architecture continuously moves between two extremes. On the one hand, the towers of Babel – usually Mies van der Rohe-style show-off buildings – and on the other, the sublime.

The latter achieve a stunning purity that derives its power from a grasp of the natural world's divine designs. Think of New York's Solomon R Guggenheim Museum by Frank Lloyd Wright, Kuala Lumpur's Petronas Towers by Cesar Pelli, and London's Swiss Re Tower by Norman Forster. They have a biotic power. That means they relate to the world of living things and their ecology. On the other hand, the genesis of the Fountain Place in Dallas by IM Pei lies in the abiotic; it is governed by the crystal. Still, all these buildings have one thing in common: they're inspired by the geometry and mathematics of natural planes and curvatures.

Now look at the cactus. Better still, close your eyes, cup your hands and approach carefully. Let your hands explore the plant. Touch the spines gently: feel their springiness, sharpness. Feel the outline: nearly always vertical or geodetic. Finger the ribs: raised and curved like cathedral buttresses. Press the trunk: firm and unyielding. Now open your eyes and examine the plant from above. It's a pure interplay of lines – the patterns of a Foucault's pendulum or a creation by Buckminster Fuller.

The cactus specialises in adversity. Where heat and drought drive all other life to shade and shelter, and all other plants give up the ghost, there the cactus flourishes. It had to develop unique structures and defences to maintain its very succulence in desolate areas. It has thorns against thirsty mouths – sharper even than the spike-thorn and flexible as a watch spring. And it is beautiful to boot. The colour of the spines is like a milky agate or the elusive yellow of a bird's beak: light shines through the edges.

The tree-like candelabra, Isolatocereus dumortieri *is a native of Mexico and a classic feature of the desert – and cartoon – landscape.*

Look again at the seams, cut deepest in the Karoo noors cactus, that poisonous candelabra of the semi-desert. These fissures and ribs ensure that the outer surface of the plant is always partly shaded and they help reduce wind currents around the plant, so that evaporation is kept to a minimum. Like an old cathedral, the structure's main supports are placed around the outside. The interior is left clear for water storage. When cacti die, they collapse like pricked balloons; all that remain are little piles of thorns.

As for the flowers: between the spines and the algae-like hairs, a silver, bullet-shaped knob will suddenly appear. Then one morning, as you walk by, there'll be a perfect flower. They look like little everlastings but the petals are soft, the colours are radiant and the stamens and pistils are prominent. Strange, that they have these burning colours, because most cacti are pollinated by moths, which are nocturnal creatures. It's a melancholy thought ... such beauty lost to the night. I have cacti on the steps of my stoep on the farm. They're grateful creatures that make no demands yet reward me daily with their heavenly shapes and designs. Get one. Plant a basilica on your balcony.

CACTUS CURIOSITIES

Here is a little brain-teaser for you: All cacti are succulents, but not all succulents are cacti. Right? Right. Second, if you manage to find a South African book about succulents in a bookshop, you deserve an award. Third, cacti, so good out there in the wild, can be little devils in the home garden. But more about that later.

It may not be love at first sight because cacti are rather strange-looking plants and appear not to have any connection to other members of the floral kingdom. Unlike ordinary plants that have stems, branches and leaves, cacti have the oddest variety of shapes and designs. There are skittles, giant clubs, soaring columns, winged trunks, broom brushes, scramblers and climbers. Then there are the typical stems that look as if they're made up of fat, round or oval discs.

Cacti originated in America, with the largest number of species occurring in Mexico. But why did they assume such extraordinary shapes? Simply by adaptation to drought: to get rid of leaves (through which moisture evaporates) and to increase their ability to store moisture inside the plants. Other plants have applied the same principles for survival, for example, aloes with their thickened leaves, and members of the *Euphorbiaceae* family with their thick, fleshy stems.

A SCULPTURE GARDEN

This amazing variety of sculptural forms can make a well planted cactus garden one of the most striking and satisfying of all. The best examples are probably to be found in America, where gardens such as Lotusland, the Huntington Botanical Gardens and the Ruth Bancroft Garden in California are magical, surreal succulent worlds. South Africa has a number of notable cactus nurseries; most of their gardens are collections of plants rather than designed gardens, but spectacular nevertheless.

Clockwise from top left:
The owl's eye pincushion, Mammilaria parkinsonii, *is a small spherical species.*
The red headed Irishman, M. pitcayensis, *has a columnar shape.*
The Mexican lime, Ferocactus pilosus, *has both orange and yellow flowered forms. (Bottom left and right.)*

CACTUS BASICS

- The greatest threat to a cactus is too much water. It is better to water them really well and then allow them to dry out completely, than to give them a little water every week – they don't like wet feet. Also resist any temptation to water cacti when you plant them. Rather leave them dry for a week.
- Find out if your cactus prefers a little shade, or would rather be in full sun.
- The best position for a cactus is a sunny, open spot facing north. Because drainage is so important, a raised bed or a rockery filled with a mix of sand, gravel and clay will make an ideal cactus garden.
- These cacti grow effortlessly: *Cephalocereus senilis* (grows to eight metres), *Cereus peruvianus* (to ten metres), *Echinocereus pectinatus* (to 30 cm with pink flowers), *Mammillaria* (to 20 cm with violet or pink flowers), and *Parodia* (to 15 cm with yellow or red flowers).

The mother-in-law's cushion, Echinocactus grusonii, *is a large spherical cactus that can grow to nearly a metre in diameter.*

The Sparklers

"Excuse me," said the old lady of Soebatsfontein, pulling up her overall to just above her knees. With teeth as strong as a mule's, she chewed flat a duiker's tendon and stripped off a bunch of thin fibres. Sticking out her lower lip, she pulled the fibres across it and then rolled the damp threads on her knee with the palm of her hand until they formed a solid strand. "This, my child, is what you use to sew a dassie kaross together."

"But how do you get the skins soft, Tannie?" She stood up stiffly, walked down the steps to the garden gate and bent down, hand on her back, to pick up a dry, twiggy stump. "You make a hollow in the sand, line it with a duiker skin and soak the furs in a strong tea of the ash of this soutslaaibos."

"Aaah," I said, "mesembryanthemum?"

"No," she replied, "soutslaaibos."

Of all our botanical curiosities, vygies are the strangest. From the purple and cream stretches of sour fig to the willow-green stone plants, vygies present a plethora of colour and form. The leaves are usually grey-green with a waxy film, almost like that of a ripe plum – touch it, and the bloom is gone. Some leaves are round, some cigar-shaped; others resemble the beak of a tortoise, smooth or flecked. And then you have the soutslaaibos, or mesembryanthemum (*mesembria* = midday; *anthemon* = flower). Its leaves are covered in sparkling papillae, hence its species name *crystallinum*.

Patiently they wait for rain, sometimes for years. When it comes, you have a vygie year. A week or so after a good downpour comes the splendid encore. Then you choose a calm, cloudless day, find a rock where you can sit, and look over the plain. Everything drab, monochrome – until midday, when suddenly the miracle occurs. All the flowers open simultaneously: before your eyes a vast ripple of colour flows from satin into pure, sparkling tints.

The flowers of mesembs shine with an unrivalled brilliance, designed to attract pollinators from afar. They vary in size and colour, and in the number and width of petals, but their structure is always the same: a daisy-like flower, with uniform petals arranged around a central boss of stamens.

I am always amazed at the riches of the West Coast. The landscape is so ancient, so bare, yet it contains a multitude of hidden treasures. Drive north from Elands Bay and you can still find horse-mills. This is bokkom country, where the local residents collect wild cabbage and waxberry honey, and they still make sour-fig jam. Have you ever tasted sour-fig jam? Go in search of it, for there are still a few people who make it. In her book *Uit die Waaisand en die Kalkklippe*, Betsie Rood explains exactly how to prepare the fruit. She ends: "Take note that between Christmas and New Year is the best time to make this jam – just when there's the least time to do it!"

My front stoep faces west and gets as hot as a blast furnace. On each step I have an earthenware pot. Each one contains a vygie, the surface of the soil covered with coffee-coloured pebbles. Every morning, I sit on the top step and drink my coffee and I watch them. They are as static as fossils the whole year long. Then suddenly, overnight, a bud will appear. That day I try to stay home and wait for the sun to strike, because I know the flower will open just for me.

Vygies give you a quiet pleasure. How sweet the old names – ghoena, ghôkum, steekvy, kougoed, vingerkanna, brakslaai, asbos, loogslaai, kirriemoer, haasballetjies. Above all, vygies reward you with their gratitude, abundance and light. At midday, in drought and grit, they shine.

MARVELLOUS MESEMBS

The word vygie is short and sweet, a drop of honey on the tongue. English has only the cumbersome Latin, *Mesembryanthemum*, usually shortened to mesemb. These wonderful succulents belong to a vast botanical family called *Aizoaceae*. There are 123 genera of mesembs, most of which occur in the southwestern regions of the African continent. They vary considerably but all have special characteristics that show Mother Nature at her most creative.

Mesembs include low-lying, twig-like shrubs, flaccid or tufted ground-huggers and thumbnail-size buttons. Leaves may be cylindrical, flat, tongue-shaped, rough, toothed, crystalline, ruffled or smooth and waxy. Sometimes the leaves are almost the entire plant, disguised as stones and invisible in their natural environment except when flowering. These flowering stones, such as *Lithops*, *Fenestraria* and *Conophytum*, are prized by collectors worldwide.

The flowers vary in size and colour, and in the number and width of petals, but they all have two things in common – they are daisy-like, with uniform petals arranged around a central boss of stamens, and they shine with a brilliance unrivalled even in the Cape floral kingdom.

This brilliance is designed to lure insects from afar, but mesembs also use additional strategies to reduce competition with other plants for pollinators. Depending on the species, flowers appear at different times of the year and open at different times of the day.

The seed capsules are marvels of engineering. Botanists have been able to divide the mesembs into six groups according to the structure and intricately varied geometry of their seed capsules. These have between four and ten seed chambers, which open in response to moisture, release seeds (which are usually washed out by raindrops), then close again as they dry out to wait for the next shower – next week or next year. This greatly enhances the chances of successful germination and species survival in arid environments.

Mesembs may be shrubby, lax and ground hugging, or fat button-like plants with flowers in a dazzling range of hot and icy colours: white, yellow, pale pink, salmon, orange, scarlet, crimson, magenta, and many delicate shadings in between.

MESEMB BASICS

- The shrubby mesembs, such as *Lampranthus*, are what most South Africans think of as vygies. They grow easily from cuttings or seeds.
- Cuttings are best taken after fruiting and started in pots in a gravelly, sandy soil mix enriched with compost and bone meal. The same soil mix is good for planting in the garden. Cuttings of the ground-cover mesembs, such as *Aptenia* and *Carpobrotus*, may be planted in situ at any time of the year provided they can be watered regularly until they are well rooted.
- Flowering stones of all kinds are best in pots.
- Plenty of sunlight (particularly in the morning) and good air circulation are important – these are plants of the wide open spaces.
- The shrubby and ground-cover mesembs are ideal for arid and waterwise gardens but not recommended for regions that are prone to frost.
- No mesemb should be overfed or overwatered. Keep fairly dry in summer.
- Plant the shrubby kind in gravel interspersed with sowings of Namaqualand annuals, for a truly brilliant spring display. Add summer- or winter-blooming aloes for structure and colour throughout the year.
- Shrubby mesembs need to be replaced with fresh plants after two or three years. These are easily grown, once again, from cuttings.

Magenta Lampranthus amoenus *is one of many shrubby mesembs which become domes or carpets of glittering blooms in spring and early summer.*

Dark Stars

"I, Carl Pehr Thunberg, a Swede, directed to the coasts of the Southern regions under circumstances as described in my *Flora Japonica*, did, during the years 1772 to 1775, with care and devotion search for, collect, examine and describe natural objects of all kinds, with special reference to the Cape Flora."

These are the words of Thunberg in his preface to the monumental *Flora Capensis*, published in 1813. He concludes, as only the explorers of old could: "… I passed through many dangers: I took care to avoid hostile tribes and savage predators and, for the sake of the beautiful plants of this southern Thule, joyfully I ran, perspired and froze."

Here, with Thunberg, we encounter for the first time the stapelia in Africa. He writes that the San (and also the early settlers) would chew the stems of the ghaap (*Hoodia*) when they found themselves without water in the veld.

The year 1908 saw the publication of Hermann Wilhelm Rudolf Marloth's *Wissenschafliche Ergebnisse der Tiefsee-Expedition auf dem Dampfer Valdivia*, in which he reports for the first time that these plants are extremely effective appetite suppressants. Over 100 years later, the pharmaceutical company Pfizer is planning to launch a new product derived from *Hoodia*, having paid $US25 million for the production patent.

The stapeliads consist of almost 20 genera, which include *Hoodia*, *Huernia*, *Tridentia*, *Orbea*, *Stapelia* and *Tricocaulon*. Africa – particularly South Africa, the Rift Valley, Madagascar and the Canary Islands – is richly endowed with succulents. This is largely owing to a long history of drought, seasonal rainfall and rocky terrain. With such conditions driving the engine of evolution, xerophytes tend to develop (*xeros* = dry, *phyton* = plant). Worldwide there are more than 20 000 drought-resistant plants, of which 10 000 are true xerophytes, and half of these occur in South Africa.

Orbea variegata *is a small speckled plant with mottled, skin-like flowers that grows on Table Mountain, and was one of the very first South African plants to be taken to Europe by plant collectors, even before Jan van Riebeeck.*

A number of strategies have given them an advantage over other plants in this pitiless environment – the most important of which are mechanical protection via thorns, chemical protection via poisonous or bitter sap in leaves and stems, superb camouflage, and the ability to grow hidden under other plants and on cliffs. South Africa has by far the greatest variety, including the stapeliads, and they are nearly all endemic, which means that they are to be found nowhere else.

It's the common name for stapelia that first catches your interest – carrion flower, a name widely used for this group of plants. Nearly all stapeliads smell absolutely awful, a nauseating, cloying stench of decay. And if you look at them against the sun, the flowers are often covered in fine silver hairs. How clever to rely on flies and bluebottles as pollinators in regions where birds are rare and unpredictable. There are yellow and black carrion flowers, but the red ones are the most beautiful. The colour is never bright, never on the surface. It appears veiled, as though under a layer of wax or powder. I think of the pinkish red of the clay that my ancestors used to paint the hearth, the misty red of hematite, creamy mauve, the rough red of rust. These colours alone are enough to captivate you, but paint them on the heavenly star shape of flowers and the combination is sublime – in spite of the smell.

Years ago, I stopped near a salt pan somewhere north of Brandvlei, in a part of the world known as Grootvloer. At a lay-by an old man, tin bucket on his arm, stood bent over a stapelia, harvesting the thorny stems. I walked over to him, unable to contain my curiosity. "Good day, Oupa, and what are you going to do with that smelly stuff?" I asked.

"Meneer," he replied, "I crush the stems, I steep them in brandy and then I drink it for my piles – and my pleasure."

And that's how I think of stapeliads. South Africa's own; helpful, delightful and unique even among our own plants. The other day I read a description in a catalogue of exotic flora: "Stapeliads, the orchids of the succulent plant world." Exactly.

AMBASSADORS EXTRAORDINARY

Whence the name? At least 20 years before the arrival of Jan van Riebeeck, the Dutch missionary Justus Heurnius stopped at the Cape on his way to the East. Here he made collections and drawings of plants, one of which, a little speckled succulent with a star-shaped flower, made its way back to Europe. It was described by the learned Johannes Bodaeus van Stapel in his 1644 edition of Theophrastus's *De Historia Plantarum*, and so the genus was named *Stapelia* in the scholar's honour. (Another genus, *Huernia*, pays tribute to the missionary's botanical zeal.) And so these faintly repellent yet mesmerising plants were among our earliest floral ambassadors.

Since then botanists have spent centuries tinkering with names and classifications. The stapeliads are now not a single genus but a group of plants that belongs to the family of *Apocynaceae* and embraces approximately 330 known species worldwide. Twenty genera, comprising 182 species, are indigenous to Southern Africa. The rest are to be found in dry regions along the southern coast of Europe as well as in Arabia and India.

Duvalia caespitosa subsp. vestitia *has the characteristic raised corona of the genus. Variations of flower form within the stapeliads are almost endless, and have provided a considerable challenge to botanists down the centuries.*

A NOVEL STRATEGY

These perennial plants grow low to the ground, with leafless, succulent stems, usually four-sided, and with notched, sometimes thorny edges. The flowers are the sensational part: evolved to attract not bees but flies as pollinators, they reveal nature at its most ingenious. The form is simple: five petals joined around the centre of the flower to create a star-like shape. They may be fleshy, hairy (like a little furry corpse), speckled (like a toad), or simply an appealing shade of liver. Instead of sweet perfume, the smell of decay lures pollinators from far and wide. They are not carnivorous flowers: the fly escapes unharmed. After pollination, many species produce a marvellous two-horned seed pod (hence another common name, bokhorinkies), which ripens to release a cloud of seeds on tiny silken parachutes.

STAPELIA BASICS

- They are not difficult to grow – provided you can keep the plants dry and warm enough.
- Find out the region of origin (summer or winter rainfall) of your species so that you can give them a dormant period without water, similar to that experienced in the wild.
- Stapeliads may be grown from seed or cuttings, which should be left to dry out for a couple of days before being planted in a well draining soil mix topped with pure sand to prevent rot. Allow the growing medium to dry out completely between waterings.
- Shallow roots make stapeliads ideal for planting in wide, shallow pots. Some species also work well in hanging baskets if they are allowed to trail over the sides.
- Species like *Stapelia hirsuta* or *S. gigantea* make pleasing ground-cover plants in drifts with other succulents.
- Stapeliads only flower on new stems, so every few years the plants need to be lifted, older stems removed and young growth replanted. Let the divisions dry out before replanting in dry soil, and water only when they start growing again.
- Protect plants from the fiercest sun and cold: in the wild they usually grow in the shelter of a rock or a bush, so they can cope with partial shade.

Most stapeliads have star-shaped flowers, but a rare few are like fat little doughnuts.
Clockwise from top left: Tromotriche engleriana; *starry yellow* Tridentia virescens; Stapelia remota; *ultra-furry* Stapelia hirsuta.

Sea Saraband

Imagine that you are watching a ballet from high above the stage. The dancers stand frozen, and then suddenly the performance begins. Viewed from above, the swoop, the arch, the undulation of classical dance is now reduced to an interplay of lines and circles; it is mathematics set to music. Only in profile or half-profile can we "read" the dance.

At the water's edge, look down into a rock pool, wait for a wave or a ripple – and watch the dance begin. No plant or animal on earth can dance like seaweed because, beneath the water, it has overcome gravity. Firmly anchored to a pivot on the rock, it performs its entrancing saraband.

The human body can also do this, with its arms and hands. View those old, grainy films of Isadora Duncan and Martha Graham and also more recent icons such as Annie Lennox. Think of Marlene Dietrich's pale hand in *The Blue Angel*, or the motherly hands of Käthe Kollwitz; think of the hands of young Balinese girls.

The thick black, lustrous hair of the Japanese can be just as fluid. They ascribe it to the regular inclusion of hijiki, a renowned seaweed of the region, in their diet. Kombu, an indispensable ingredient in soups, stews and rice dishes, is regarded as equally wholesome. In Valdivia, Chile, the Mapuche Indians still harvest and dry seaweed for the pot just as Darwin described in his *Voyage of the Beagle*. The Irish, Scots and the native tribes of North America all have traditions involving the use of seaweed and kelp forests. How odd, then, that we in Africa appear to have almost no seaweed lore.

Under seafood we should include seaweed. This is something we tend to do far too little. "It's a shame," the old tannie said to me, "that people no longer know how to eat seaweed creepers!" In *Polfyntjies vir die Proe* (Tafelberg, 1963), poet and Cape food historian C. Louis Leipoldt called it seaweed creeper. Author Jan Rabie called it jelly weed. It was with Jan that I experienced South Africa's edible seaweed for the first time when a friend and I looked him up in Vermont.

Rich in minerals and trace elements, seaweed is highly nutritious and features in numerous dishes in Asian cuisine. This bright green sea lettuce, a species of Ulva, *is often eaten in the Far East as an ingredient in salads.*

We found him on the rocks, coming home with a bag of periwinkles. He had also collected the wax-white weed on the bleached pebbles where winter storms had tossed up a flotsam of sea bamboo.

Once home, he washed and soaked it, cooked it slowly and then strained it through a cloth. He then stirred in sugar, sherry, cinnamon, cloves and lemon peel and poured the mixture into moulds. Dessert that day was sweet and sour, tart and soothing – with a whisper of the sea at the back of the tongue.

Seaweeds are unique plants, so unique in fact that they have an order all to themselves: *Chromista*. Chroma?

The colour of sea grass and seaweed is limited to brown, red and green, but brown dominates. Does this make the underwater aquarelle monotonous? A hundred times, no. Cecil Higgs spent a lifetime painting rock pools and gave new definition to the underwater palette. Glowing pink and brown together? Yes. At low tide, look at the colour gradations where ocean plants are drying out against the rocks. Forget about design, line and form.

Succumb to tint and texture – and pure abstraction.

THE GARDENS OF THE SEA

Seaweeds are not strictly plants but rather marine algae. They have no roots because they absorb nutrients directly from the sea and they have no rigid stems because they are supported by water. This flexibility enables them to survive tides, waves and currents. They are not only crucial in the marine and terrestrial food chain but also play a vital role in enriching the earth's atmosphere with oxygen.

Because they need light to survive, seaweeds are found only in the relatively shallow parts of the oceans, usually in depths from eight to 35 metres. Some species may be found down to depths of 250 metres in waters that are exceptionally clear.

Others grow in the zone exposed at low tide and are specially adapted to high saline conditions and periodic drying. All these seaweeds are firmly anchored to a solid object, usually rock. Many are equipped with tiny bladders that buoy them, to bring them as close to the sunlight as possible. A number of free-floating species, such as sargassum, are also equipped with bladders.

There are three recognised groups of seaweeds, based on their pigments. They absorb light of particular wavelengths, which gives them their characteristic colours of green, brown or red. (These colours may change dramatically once out of the water.) The seaweeds occur in a variety of shapes and sizes, from large kelps that form sea forests, to encrusting corallines. Some, especially the larger reds, are showy and attractive, while others may be inconspicuous and grow in a low "turf" on the rocks.

There are five to six thousand species of marine algae worldwide, over 700 of which are to be found around the coasts of Southern Africa. Some of these are only found on the warmer east coast, while others are found only along the chillier south. Like so much of our terrestrial flora, many of them are endemic to our region.

Clockwise from top left:
Laminaria pallida *grows to about 10 metres in length, in waters up to 30 metres deep, and is often found washed up on the beach. It has a single broad, fan-shaped blade that becomes irregularly split, hence its name, the split-fan kelp.*
Plocamium corallorhiza *is found along both the Cape and KwaZulu-Natal coasts.*
Cladophora capensis, *a delicate green seaweed, may be seen exposed on rocks at low tide around the Cape Peninsula.*
Look for the broad blades of Schizymenia obovata *on subtidal rocks.*

SEAWEED BASICS

- Seaweed is packed with nutrients, minerals and trace elements, and has been used as food and fertiliser for aeons. Today, extracts are used in thousands of products, from food additives and vitamin supplements to all kinds of chemical products.
- The ingredient that gives seaweed its unique, gelatinous quality is a pectin known as agar-agar. It is useful for making desserts and for thickening soups and stews. "All fish soups are improved by boiling some green seaweed with them," according to Leipoldt.
- The local seaweed industry is worth about R15 million annually, and has the potential to expand, both by greater use of existing natural resources and by farming.
- In the Far East, the nori (the red seaweed *Porphyra*) industry is worth $US1,5 billion annually. That's more than three times the entire South African fishing industry.
- No seaweed is known to be poisonous.
- Local interest in seaweed as a food was recently revived by a series of articles on edible seaweeds written by artist Louie Lemmer and published in the magazine *Village Life*. Louie also learnt about seaweed from Jan Rabie and developed her recipes over many years. At the time of her death, in February 2007, she had been planning a book. Back copies of *Village Life* contain her articles. For more information go to www.villagelife.co.za.

Codium extricatum *is a green seaweed common in tidal pools.*

Plants of the Dream Time

Henri Rousseau, painter of fertile imagination and compulsive visitor to the Jardin des Plantes in Paris, once told the art critic Arsène Alexandre: "When I enter the glasshouse and look at the strange plants from exotic lands, it seems to me that I enter into a dream ..." And indeed it was a dream.

When you look at "Woman walking in an exotic forest" or my favourite painting, "The hungry lion leaps upon the antelope", you can see that every aspect seems removed from reality, except for the ferns – these are rendered true to life. Rousseau simply could not reduce them any further. Simple, flowerless and hermaphrodite, they are our links with the dream, and thus with prehistory.

Ferns, together with the gingko, the welwitschia and cycads, are the oldest and most primitive plants on earth. Unlike the more advanced vascular plants, they are independent opportunists that reproduce without seed; they are tougher than they seem and occur all over the world. We think of ferns as the understorey of a tropical jungle, shadowy in the half-light, while drops of water hang at the tip of each frond.

Nothing could be further from the truth, for ferns grow in the most inhospitable of places, even right at the edge of the Namib desert.

The most striking aspect about ferns is their architecture, in particular the larger species and the tree fern. If you look at these plants from above, you will notice the perfect symmetry of the leaf, so tough it seems made of foil. Towards the centre, the rosette becomes brown and hairy – a knob of fronds about to unfurl.

Look again in profile, and the most beautiful element of the plant becomes apparent. Like the elegant neck of a violin, or the curved hand of a dancer, the unfurling fronds rise from the fountain of leaves.

A large crown of glossy fronds makes the dwarf tree fern, Blechnum gibbum, *a showy plant for indoors or out, provided it can be sheltered from frost. This species has a rhizome which develops into a slender trunk that can grow to nearly a metre tall with age.*

A lovely fable about the fern comes from Finland: if, on midsummer's eve you should find the seed of a fern in full bloom, you would receive the cloak of invisibility and be led to the ignis fatuus or will-o'-the-wisp, which burns eternally bright over caches of hidden treasure.

Both the San peoples and Australia's Aborigines are still able to enter dream time through song and ritual. In the course of millennia, these rituals and symbols have remained unchanged as vehicles of the imagination.

Plant your ferns with care, where you can see them at night. Seat yourself nearby and perhaps, late one night, you will suddenly glimpse one of Rousseau's fables, peeping at you through the fronds.

JUST FRONDS

Ferns evolved in a time when there were no pollinating insects, hence their aloof independence as hermaphrodites. They have no flowers, no fruit, only cool green frond upon frond, pattern upon pattern, and dust-like spores which are released from those brown dots along the underside of the mature frond – the beginning of a highly complex reproductive process.

There are approximately 12 000 species of fern worldwide, about 260 of which are indigenous to Southern Africa. They grow in a wide variety of climates, but are usually restricted to a few specific habitats. Typical of these are acid wetlands, areas of mist or high rainfall and damp subtropical forests, where some are forest-floor plants and others epiphytes growing on trees. In dry climates and open conditions, look out for ferns in rock crevices, seepages and stream banks, or anywhere shaded from the sun for at least part of the day.

There are the tough, dry ferns found on koppies, the amazing blue rock fern, *Pellaea calomelanos* with its black stems, and the lip fern *Cheilanthes hirta*, delightful surprises in dry places. And then there are the ferns of the Cape forests and the kloofs and waterfalls of the eastern escarpments, worlds of enchantment. In Mpumalanga and KwaZulu Natal in particular, the habitat of mountain and forest ferns has been drastically reduced by commercial forestry and they now survive in numbers and true environment only in the more inaccessible patches of natural forest, kloofs and crags.

All indigenous ferns are protected and they may not be removed from the wild without the necessary permit.

THE FERN GARDEN

In these times of global warming, every waterwise garden and certainly every Karoo garden should consider converting a shady corner into a small fernery. Cover the walls with ivy and *Asparagus plumosus*. Go for as many indigenous ferns as you can, and put them on stands, tree stumps and in hanging baskets. An exotic staghorn fern is a must for atmosphere, if you can find one. Add forest plants like *Streptocarpus*, *Haemanthus* and *Scadoxus*, ensure a trickle of water and one chair: rehydration for the parched soul.

The bird's nest fern, Asplenium nidus, *is native to the jungles of Asia and northern Australia and may be terrestrial or epiphytic in its natural habitat. It needs warmth and plenty of moisture and humidity to grow well.*

FERN BASICS

- Ferns require shade for all or part of the day. Some are adapted to deep shade but most prefer it light or dappled.
- The soil should be kept damp.
- Humidity is important. Use a sprinkler or micro-irrigation every day or give your ferns a mist rain, especially if they are planted in containers indoors.
- A sandy soil with plenty of fine compost, such as can be found under thorn trees, is ideal. Ferns mostly prefer a slightly acid soil with a pH of 5,5 to 7,5.
- Temperatures should not be too high or too low – a safe range is 5°C to 35°C.
- Certain exotic species, like the sword fern *Nephrolepis exaltata* and the *Azolla* species of water fern are declared weeds in South Africa.

Clockwise from top left:
The Tasmanian tree fern, Dicksonia antarctica, *is one of the most popular landscaping plants around the world.*
The silver lace fern, Pteris ensiformis *"Victoriae", is an indoor favourite.*
The dwarf tree fern, Blechnum gibbum, *has a fishbone frond typical of many ferns.*
Toothed brake, Pteris dentata, *is a lovely indigenous fern for semi-shade.*

Sylphs of the Night

Kampala knows no seasons. Lying almost on the equator, the city is like an aquarium – dank, green and oppressive – because Lake Victoria will never allow you to forget her presence. The night air is charged, stifling and full of sounds. A saffron-yellow moon rises above the low hills to shine in at your window; a hawk moth taps against the mosquito gauze. You sleep fitfully. Finally, drenched in perspiration, you get up around midnight and go outside, down the street past the Stanley Hotel and into the city.

Here you need to tread carefully because manhole covers have been stolen, open drains cut across the pavements and there are virtually no streetlights. All around there are frogs calling, unfamiliar insects shrilling and marabou storks scuffling about in their untidy nests.

In the open darkness of a park you walk into a cloud of perfume – sweet, sweet, sweet – with the creamy undertone of clean skin after a bath, hanging in the air like a bank of mist. On the far side a wall of moonflowers gleams – sylphs of the night. Only then do you realise that Kampala nights have always been perfumed with moonflower, and this is the most likely source of your tropical enchantment and unease.

Is there another plant that blossoms as frequently in art, music and literature? Literary giants such as Faulkner, Theroux, Thoreau and Steinbeck have used moonflowers in their stories and novels. Léo Delibes' Lakmé dies after eating moonflower leaves. Tori Amos composed ecstatic and hypnotic rhythms around the plant and, in modern music, especially in the dark sounds of the Goths, moonflowers feature again and again.

The finest work of that troubled Mexican angel, Carlos Castaneda, was written in the shade of the moonflower.

I can gaze for hours at the architecture of the trumpets. The curves of the edges are as delicate

The moonflower structure is rather like an umbrella, with five or more delicate ribs that are wrapped in a spiral in bud, unfurling to release an exquisite perfume. Trumpets may be anything from 10 cm to 25 cm, depending on the cultivar.

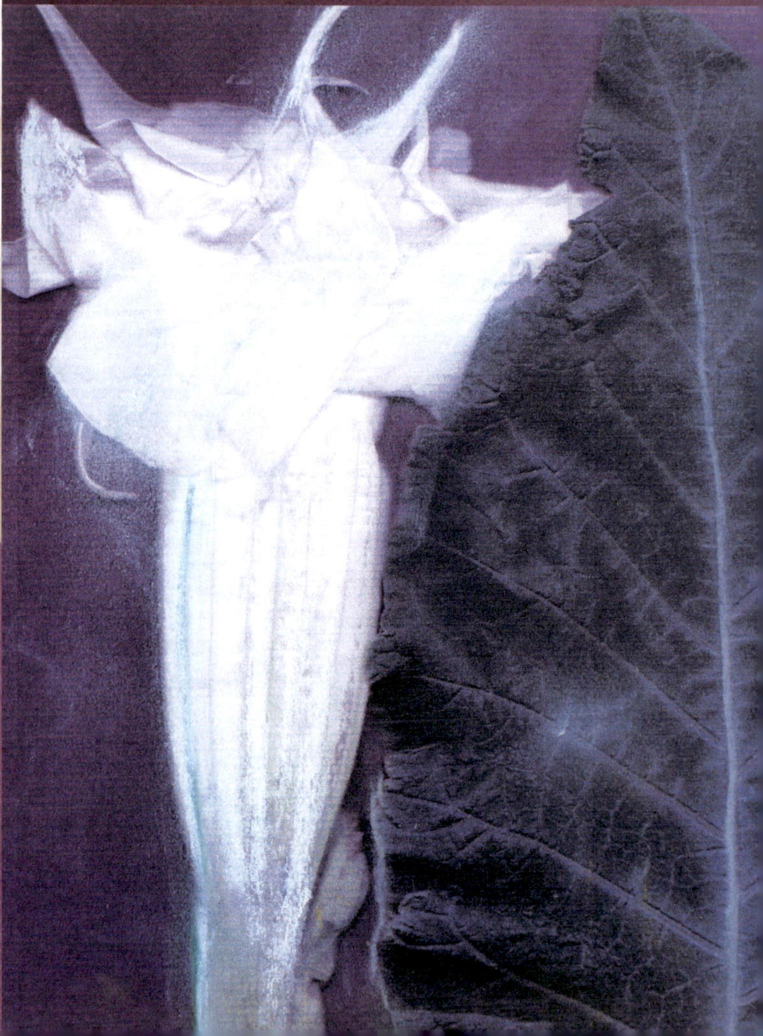

as the hem of an Erté dress, the colours ephemeral. Kampala's blooms are all white, but nowadays there are cultivars of pink, yellow, mauve and even gold.

Garden flowers such as asters, roses and bougainvilleas have fixed colours, unchanging no matter the light. Moonflowers are opalescent because the trumpet is all structure, like a little Modernist building. The ribs are flying buttresses to hold the flower open, bringing the colour to life like a dragonfly's wing. Viewed from below it becomes a mandala, as in Georgia O'Keeffe's "Jimson weed", painted in 1932.

Plant a pure white moonflower in your garden and wait for a perfectly still summer's evening. Like a succubus, the trumpets, glowing like those pale Edwardian lamps of ribbed glass, will draw you closer. The perfume of the flowers reaches you mixed with the rank smell of a crushed leaf. Ponder them late into the night – angel's trumpets, devil's trumpets, thorn apples – or *malpitte* (seeds of madness)?

From Joaquín Ramón Martínez Sabina's *A Song for the Magdalene*:

If, in your glove compartment
You have hidden your lost soul
Park your car next to her incandescence
Her abundance of milk and honey
You will be thrown a curve,
A curve of unexpected redemption
The most forbidden of all fruits
Waits for you till dawn ...
(Translated by Joanna Harcourt-Smith)

HANDLE WITH CARE

The *Datura* genus is probably native to the Americas, but a long history of medicinal and ritual use has led to its distribution and naturalisation worldwide.

The genus embraces an uncertain number of species and originally included both declared weeds and prized garden plants. All contain powerful tropane alkaloids, making them poisonous if eaten. These alkaloids imbue the plants with valuable medicinal properties and rather less useful, even fatal, hallucinogenic ones.

Legend has it that datura seeds or leaves were used by the priestesses of Ancient Greece to induce the trances and fits suggestive of divine power, and the list of more recent recorded fatalities and permanent derangements is a long one – hence the Afrikaans name, *malpit*. The alkaloids are, however, used commercially to beneficial effect in medications for Parkinson's disease, asthma, in eye drops and various other external applications.

In South Africa, *Datura stramonium* is widely used in traditional medicine to relieve asthma and reduce pain. Fresh leaves may be warmed and used as a poultice, or dried and smoked for chest complaints. The green fruit is applied locally for toothache and tonsillitis.

Clockwise from top left:
Withered flowers reveal the delicacy of the fabric.
The velvety leaves are a perfect foil for the translucent delicacy of the flower trumpets.
Moonflowers are creatures of the night, aptly named.
The rib of each petal ends in a fine flourish.

MAGIC IN THE GARDEN

Most of the weedy species, such as *Datura stramonium*, are annuals that bear their flowers at various angles on the plant. The garden species and cultivars are tall, woody shrubs with soft velvety leaves and may grow into small trees in favourable positions. Though often still referred to as *Datura*, these handsome plants have been reclassified as *Brugmansia*. Their graceful pendant trumpets, which may be up to 25 cm long, also distinguish them from the annual species. Fragrance aside, these sumptuous curtains of flowers in white, cream, yellow, gold, apricot, pink and mauve, lend brugmansias a gorgeous visual presence.

Every garden should find a place for at least one. It's merely a question of choosing the right colour to suit your style, because the form of this beautiful plant seems to work with all kinds of plantings in all kinds of settings – indigenous, formal, cottage, and tropical, of course.

Most brugmansias available from nurseries are variants of *Brugmansia x candida*. The double white is breathtakingly elegant as well as superbly fragrant. Golden apricot "Grand Marnier" is a dazzling tour de force, while the pink and salmon varieties offer a more delicate enchantment. *B. sanguinea* is a species with more tubular, red or orange flowers lit with yellow – another tropical whimsy.

MOONFLOWER BASICS

- Moonflowers are easy to grow, provided they can be protected from frost and wind.
- All you need is a semi-hardwood heel cutting or a hardwood cutting, best taken in late summer or autumn. Once the cuttings have rooted, plant them out in plenty of compost, and water them regularly and well.
- Be sure to plant moonflowers where you can enjoy the fragrance of an evening.
- Plants usually grow to three metres, or to five metres in favourable conditions, with a spread of 1,5 metres to two metres.
- They respond well to cutting back but probably the best procedure is to prune regularly to shape them. Care must be taken to maintain an attractive branch pattern for the plant to look its opulent best.
- Moonflowers also make excellent pot subjects for patios and courtyards but beware of hot, dry situations against walls, where they could be prone to red spider mite.

A trumpet within a trumpet. The double forms of moonflowers are perhaps even more beautiful than the single blooms. Flowers may be cream, yellow, gold, apricot or pink, but white remains a favourite.

Bibliography

Bown, D. 2002. *The Royal Horticultural Society Encyclopaedia of Herbs and Their Uses.* London: Dorling Kindersley.

Brown, N., Jamieson, H. & Botha, P. 1998. *Grow Restios.* Cape Town: National Botanical Institute.

Bruyns, P.V. 1996. *Stapeliads of Southern Africa and Madagascar.* Hatfield: Umdaus Press.

Burr, C. 2002. *The Emperor of Scent.* London: Arrow Books.

Coats Palgrave, K. 1983. *Trees of Southern Africa.* Cape Town: Struik.

Coombes, A.J. 1985. *Dictionary of Plant Names.* Portland: Timber Press.

Du Plessis, N. & Duncan, G. 1989. *Bulbous Plants of Southern Africa: A Guide to Their Cultivation and Propagation.* Cape Town: Tafelberg.

Glendinning, V. 1988. *Vita: The Life of V. Sackville West.* London: Phoenix.

Goldblatt, P.A. 1986. *The Genus Watsonia.* Cape Town: National Botanical Gardens.

http://en.wikipedia.org/wiki/Fern

http://en.wikipedia.org/wiki/Henri_Rousseau

http://nowsmellthis.blogharbor.com

Kingsolver, B. 2007. *Animal, Vegetable, Miracle: A Year of Food Life.* New York: Harper.

Larkcom, J. 1995. *Vegetables for Small Gardens.* London: Hamlyn.

Leistner, O.A. 1990. *Grasses of Southern Africa.* Pretoria: National Botanical Research Institute.

Lindler, H.P. & Kurzweil, H. 1999. *Orchids of Southern Africa.* Rotterdam: A.A. Balkema.

Pienaar, K. 1993. *Grow South African Plants.* Cape Town: Struik.

Pringle, E.L.L., Henning, G.A. & Ball, J.B. 1994. *Pennington's Butterflies of Southern Africa.* Cape Town: Struik Winchester.

Reynolds, G.W. 1969. *The Aloes of South Africa.* Cape Town: Balkema.

Rood, B. 2001. *Uit die Waaisand en die Kalkklippe.* Cape Town: Tafelberg.

Rourke, J.P. 1972. *Taxonomic Studies on Leucospermum.* Cape Town: National Botanical Gardens.

Smith, C.A. 1966. *Common Names of South African Plants. Memoirs of the Botanical Survey of South Africa No. 35.* Pretoria: Government Printer.

Smith, G. et al. 1998. *Mesembs of the World.* Pretoria: Briza.

Van Jaarsveld, E. 2005. *The Southern African Plectranthus.* Cape Town: Fernwood.

Van Jaarsveld, E. 2000. *Indigenous Waterwise Gardening.* Cape Town: Tafelberg.

Van Laren, A.J. 1935. *Cactus.* Los Angeles: Abbey San Encino Press.

Van Oudtshoorn, F. 1992. *Guide to Grasses of South Africa.* Pretoria: Government Printer.

Van Wyk, B.E. & Gericke, N. 2000. *People's Plants.* Pretoria: Briza.

Van Wyk, B-E. & Smith, G.F. 2008. *Aloes in Southern Africa.* Cape Town: Struik.

Van Wyk, B-E. & Smith, G.F. 1996. *Guide to the Aloes of South Africa.* Pretoria: Briza.

Van Wyk, B., Van Oudtshoorn, B. & Gericke, N. 2002. *Medicinal Plants of South Africa.* Pretoria: Briza.

Von Breitenbach, F. 1987. *National List of Indigenous Trees.* Cape Town: Dendrological Foundation.

Your Gardening Questions Answered. 1989. Cape Town: Reader's Digest.

www.foodgardensfoundation.org.za

www.plantzafrica.com

www.plantzafrica.com/medmonographs

www.succulents.net

Index

Page numbers in bold refer to photographs